Timbering For Small Underground Mines

Idaho Bureau of Mines Bulletin 21

*Compiled by the staff of
Idaho Bureau of Mines and Geology*

with an introduction by Kerby Jackson

*This work contains material that was originally published by
the State of Idaho with the assistance of the United States Geological Survey.*

*This publication was created and published for the public benefit,
utilizing public funding and is within the Public Domain.*

*This edition is reprinted for educational purposes
and in accordance with all applicable Federal Laws.*

Introduction Copyright 2014 by Kerby Jackson

Introduction

It has been over fifty years since the Idaho Bureau of Miner released its Bulletin 21, which was otherwise known as "Timbering and Support For Underground Workings For Small Mines".

The author of this work, the late William Wesley Staley, was well known in those days for his expertise and his contributions to the mining industry, especially in Idaho. With his background as a mining engineer, he was especially suited to write a volume such as this one, and his concise and often straight forward writing style was especially appealing to information hungry small miners of his era. Not only were Staley's numerous publications on mining topics much sought after by miners who knew about his works, but also, many of his fans attributed much of their success at mining to him.

W.W. Staley was born on September 24th, 1898 in Singers Glen, Virginia. Despite this, he actually grew up in Portland, Oregon where he attended local schools. Though little is known of his earliest years, it is known that after graduating high school, Staley spent several years working as a chemist at a Portland cement plant.

In 1921, Staley enrolled at the New Mexico School of Mines and graduated in 1925 with a bachelor's degree in Mining Engineering. Not yet satisfied with his education, by 1931 he had also received a master's of science degree in metallurgy from the University of Idaho and a professional degree as an Engineer of Mines from the New Mexico School of Mines.

In 1930, William Staley joined the staff of the University of Idaho's School of Mines where he began to conduct research in mining and metallurgy. Up until his retirement in 1968, Staley authored numerous academic articles, textbooks and industry reference books on mining related subjects.

His academic work was so well received by the mining industry that he also acted as a consultant to some of the leading lead, silver and zinc mining companies in the United States, as well as worked as a consultant for the Atomic Energy Commission and the U.S. Bureau of Indian Affairs in the processing of mining claims held by several Indian tribes. Following his retirement from the University of Idaho in 1968, Staley moved to Tucson, Arizona where he split his time between consulting to the mining industry in that state and holding seminars on earth sciences at the University of Arizona until his death at the age of 93 in 1992.

In addition to this volume, some of William Staley's written contributions to the mining industry include such titles as "Mine Plant Design" (1936), "Introduction to Mine Surveying" (1939), "The Design of Small Wooden Headframes" (1937), "Prospecting and Developing A Small Mine" (1961), "Gold In Idaho" (1960), "Elementary Methods of Placer Mining" (1944) and twenty or more some others.

William Staley's written works, though now mostly retired to the dust bin of history, remain important because in large part, he wrote them with the small independent miner in mind and seemed acutely aware of his audience's often limited amount of engineering experience and capital. As a result, Staley's written works were of practical use to the small miner and the measure of their value seems only to increase as time goes on and this type of first hand knowledge becomes lost from generation to generation.

Despite the fact that we live in the so-called "Information Age" where information is supposedly only the push of a button on a keyboard away, true insight into age old mining techniques are illusive and often hard to come by, even to those of us who seek out this sort of information as if our lives depend upon it. Without this type of information readily available again, there is little hope that the average independent miner will ever have much of a shot at success.

This important volume and others like it, are being presented in their entirety again, in the hope that the average prospector will no longer stumble through the overgrown hills and the tailing strewn creeks without being well informed enough to have a chance to succeed at his ventures.

Kerby Jackson
Josephine County, Oregon
January 2014

TABLE OF CONTENTS

	Page
ABSTRACT	1
INTRODUCTION	2
ACKNOWLEDGMENTS	3
ESTIMATION OF GROUND PRESSURE WITH SELECTION OF SUPPORT	4
Selection of cap	9
Selection of posts	11
Post in bending	12
Force acting on spreader	13
Suggested changes in working stress	14
TREATMENT OF TIMBER	17
FRAMING OF TIMBER	18
INSTALLATION OF TIMBER	19
TIMBER REPAIRING	20
DISCUSSION OF ILLUSTRATIONS	21
Simple drift supports	21
Headboards	21
Squeeze sets	22
Erection of drift sets	22
Chute construction	22
Shaft bearing sets	23
Cribbing	23
Application of stulls	23
Loose or squeezing ground	23
Vertical and inclined shafts	23
Square-set timbering	24
Application of rock bolts	25
Bolting patterns	26
Beams and roof ties	26
Anchorage capacity and bolt loading	27
Problems of bolt use	28
Advantage of roof bolts	29
Installation sequence	30
Slotted type of roof bolt	30
Headed type of bolt with expansion shell	30
Use of steel sets	31
Use of concrete sets	31
Design of concrete set	33
Ordinary set	34
Loaded drift set	42
Calculation of mix	46
Placing of sets in drift	49
Chute set	49

Side lagging........	49
Wooden headframes ...	49
Wooden ore bins	50
REFERENCES CITED.....	53

TABLE OF ILLUSTRATIONS

DRAWINGS

(Facing pages indicated except as shown)

Figure	1	Simple stull support..	7
		(a) In the vein with angle of underlie...................	7
		(b) Outside the vein...	7
Figure	2	Drift set..	7
Figure	3	Headboards...	8
		(a) Squeeze heading..	8
		(b) Plain heading...	8
Figure	4	Drift sets...	9
		(a) Unframed..	9
		(b) Framed..	9
Figure	5	Effect of heavy ground...	10
		(a) Before compression of heading......................	10
		(b) After compression of heading........................	10
		(c) Cap failure...	10
		(d) Cap failure...	10
		(e) Loading for designing drift set......................	10
Figure	6	Lagging...	11
Figure	7	Drift sets below stopes..	11
		(a) Framed pony set..	11
		(b) Unframed squeeze set....................................	11
Figure	8	Standing posts and cap from staging.......................	12
Figure	9	Placing cap from muck pile.....................................	12
Figure	10	Placing cap in loose ground with boom...................	13
Figure	11	Chute construction..	13
		(a) In timbered drift..	18
		(b) Construction detail..	14
		(c) In untimbered drift..	14
		(d) In untimbered drift..	following page 14
Figure	12	Bearing sets - shaft...	" " "
		(a) Wall plate bearing...	" " "
		(b) End plate bearing..	" " "
Figure	13	Ladder...	" " "
Figure	14	Wedge...	" " "
Figure	15	Stulled raises...	16
Figure	16	Raise timbering..	16
Figure	17	Cribbing..	17
		(a) Mortised one side, tight lagging....................	17
		(b) Mortised both sides, tight lagging.................	17
		(c) Mortised both sides, space between lagging..	17
		(d) Unmortised, round lagging............................	17

Figure	18	Use of stulls	18
		(a) Vertical, headboard and footboard	18
		(b) In inclined stope, angle of underlie	18
		(c) Leveling up stull in inclined vein	18
		(d) Catching up caved area	18
Figure	19	Forepoling (spiling)	19
Figure	20	Shaft set for vertical shaft	20
Figure	21	Shaft set for inclined shaft	21
Figure	22	Square-sets	22
		(a) Cap butting	22
		(b) Post butting	22
Figure	23	Rock bolts	26
		(a) Wedge-type	26
		(b) Expansion shell	26
		(c) Wooden bolt	26
		(d) PERFO (mortar)-type	26
Figure	24	Application of rock bolts	27
Figure	25	Steel sets	28
		(a) Yieldable arch	28
		(b) Ordinary steel set	28
Figure	26	Concrete set design	following page 33
		(a) Top of web bar	" " "
		(b) End anchorage	" " "
		(c) Cross section of cap	" " "
		(d) Post and cap	" " "
		(e) Cross section of post	" " "
		(f) Cross section of cap	" " "
		(g) Cap	" " "
		(h) Cross section of post	" " "
		(i) Shear diagram	" " "
		(j) Post	" " "
		(k) Water-cement diagram	" " "
		(l) Timber facing on post for chute	" " "
Figure	27	Wooden headframe	49
		(a) Four-front posts	49
		(b) Two-front posts	49
		(c) Strap-bolt detail	49
Figure	28	Wooden ore bin	50

TABLES

Table	1	Estimated rock load	5
Table	2	Allowable stresses in mine timbers	6
Table	3	Height of rock	8
Table	4	Strength of concrete	35
Table	5	Recommended slump for various uses of concrete	35
Table	6	Areas, perimeters, and weights of standard bars	35
Table	7	Trial mixes with medium sand and 3-in. slump	36
Table	8	Essential details for various loads on wooden headframes	51

TIMBERING AND SUPPORT OF UNDERGROUND WORKINGS FOR SMALL MINES

by

W. W. Staley

ABSTRACT

No matter how limited their extent, few underground mining operations progress very far without requiring some sort of support. This Bulletin presents discussion and sketches as a guide for designing underground support of timber, concrete, or steel in relatively shallow workings.

Because timber will satisfy most support requirements, this discussion offers methods for supporting drifts, raises, shafts, and stopes, as well as suggestions for framing and later transporting packaged units into the mine. And because small mine operators may not be aware of the importance of properly blocking timber sets in place, this frequently slighted but important aspect of the supporting process is discussed. Drawings and sketches are used to show various timber sets and methods of erecting and blocking them and the accompanying text includes pertinent supplementary remarks. Another subject introduced is treatment of timber for prolonging its life, which, on occasion, should be considered even by the modest operator. Then, because rock bolts are in wide use today for replacing or supplementing timber, their use and installation are extensively outlined. Also the advantage of replacing collapsed headboards or blocking before failure of the main support is discussed. Several drawings and designs are offered for actual dimensioning and design of chutes for limited tonnage operations, which has been inadequately presented in the past.

For areas where timber is scarce or expensive, reinforced concrete sets of simple design should be substituted. Designing these sets by ultimate strength procedure is discussed in some detail. Similarly, a higher working stress for timber is also suggested, as are standards in working stresses for concrete, steel, and timber in mine-support design that are different from those used for surface structures.

Although headframes and ore bins are not a part of mine support, most mine plants find them necessary; for this reason information on these two surface structures is also included.

INTRODUCTION

Support of underground openings is much the same for both large and small mining operations. Essentially, the difference lies in the depth and in the extent to which the two types of work are compared.

Most large mines are deep, with many miles of drifts, raises, stopes, and shafts with open spaces more or less loosely supported by rock pillars, fill, and timber or other material, or by a combination of such supports. As a rule, the difficulty of maintaining openings increases with depth, but even shallow workings at times encounter very heavy ground and may also be disturbed by rock bursts.

Prompt consideration should be given to one of the leading causes of ground failure: the effect of moist air on certain rock-forming minerals. Many subsequent support problems could doubtless be minimized if the circulation and penetration of moist air were immediately controlled.

An explanation of ground support is best presented by illustrations supplemented with written explanation. In the present discussion, strict "drafting rules" have not always been followed when preparing the illustrations. In many instances, where adhering precisely to rules of mechanical drawing would have caused confusion, liberties have been taken with common drafting practice, and certain irrelevant lines have been omitted.

Every effort should be made to standardize design and dimensions of timber sets. Also, the number of saw passes and cuts during framing should be reduced to the minimum.

Standardizing drift sets, chute assemblies, raise sets, and other applications reduces both the timber inventory and the confusion of having several pieces serve the same purpose. In some instances pieces of a set may be made interchangeable (for example, caps and girts in square-sets).

When it is warranted, similar pieces may be packaged by using wire wrapping or iron strapping. Wedges, chute assemblies, and other units should be packaged for transportation underground.

Even a casual inspection of mining operations will indicate many applications of timber in addition to those presented here. Such additional uses are left to the ingenuity as well as the requirements of the reader.

A discussion of ground support would not be complete without including at least an introduction to rock bolts, steel sets, and concrete sets. Occasional use of these materials, especially rock bolts, is not unusual at small mines. Concrete sets may be especially useful where localities are not well supplied with timber, or where underground conditions promote rapid decay of timber. Under these circumstances, concrete might be used economically by the small mine operator.

ACKNOWLEDGMENTS

It is impossible to give credit for most of the material included here. Among miners and supervisors the bulk of this information is passed on orally to the younger generation. Hence, the exact origin of timbering procedures is indeterminable. Original ideas have been modified (many times without improvement) from camp to camp. Not too many years ago, when miners were a constantly migrating crew, an original idea developed in any given district could have been introduced in an astonishingly short time (almost simultaneously) in mining camps all over the world without credit to its originator.

In those few instances where such information is known, I have credited the source; but in most cases, however, I must simply acknowledge gratefully the contributions of those first unknown miners and timbermen.

ESTIMATION OF GROUND PRESSURE WITH SELECTION OF SUPPORT

Ordinarily, one must know rock mechanics to estimate ground loads. For the underground conditions usually encountered in shallow- to medium-depth operations, however, a detailed application of rock mechanics is of doubtful value. Certain maximum conditions may be assumed. As a rule, underground support based upon such assumptions will be quite satisfactory, at least for temporary service.

As a basis for estimating loads on timber sets (loads on steel or concrete may be estimated in a similar way), the following data are given.

Table 1 (Proctor and White, 1946, p. 91 and Feather River and Delta Diversion Projects, 1959, p. 67) gives information for 10 different ground conditions.

The material in Table 1 may be better understood by consulting Figure 5e. According to Proctor and White, the load formulas have been amply verified by experiment and practical observation. A few remarks will explain the data. Column 1 classifies the ground condition by using the physical properties and mineralogical composition of the rock. The 10 conditions shown will apply to any type of ground. Column 2 gives the formula for calculating the maximum height, H_p, of loose rock; in other words, the height at which a permanent arch is likely to form. (For want of a better term, arch is used, although grave doubts exist as to whether a true arch actually forms). And finally, Column 3 suggests a spacing for the sets.

In Figure 5e, the _average_ side pressure on the drift is given by $p_h = 0.3w(0.5H_t + H_p)$ pounds per square foot, where w is the weight of a cubic foot of freely flowing, sandy material (Proctor and White, 1946, p. 63). The pressure is considered to result from material with a consistency of and ability to flow like sand. This analogy is justifiable because certain rocks will, upon alteration, form a sand-like mass (for example, certain granites, gneisses, schists). A difficult decision arises when assigning a value to w.

Note well the units used for p_h. When the formula is solved, the answer is the _average_ _pressure_ in pounds per square foot, as will be illustrated later.

Figure 5e suggests that the load on the cap results from combining two different portions of the loose rock mass. One of these, W_A, will depend for its bulk on the point at which arching (stabilization) begins. This zone is rectangular in section (section A). Above section A the loose mass will assume a more or less irregular shape, depending on the physical properties of the rock. For the purpose of the present discussion the outline representing the arch will be assumed to be either W_B or W_C. The total load W on the cap will result from $W_A + W_B$ or $W_A + W_C$. Blocking of the cap is assumed to be so tight that the cap will, at least when first installed, support W without help from the posts.

This whole problem is based on rather uncertain data; therefore, the weight of the cap is ignored. It would be negligible compared to the weight of rock.

Table 2 (Wood Handbook, 1935, p. 50-53 and 105 and Staley, 1949, p. 51) gives working and ultimate stresses for timber species commonly available for mine use. The working stresses listed in the table have been adjusted so that they can be applied to material similar to common grade and probably somewhat wet. Most mine timber will have been treated for protection against decay. The values given are conservative.

Column 5 gives the straightline formulas for columns (this is the type of loading that the posts in a tunnel or drift set would resist; some caps may be loaded in a similar fashion). Straightline column formulas are easily applied and are sufficiently accurate for approximating the solution of ground-support problems.

Table 1 -- Estimated Rock Load (See Figure 5e)
Rock load H_p in feet of rock on roof of support in tunnel of width B and height H_t at a depth greater than 1.5 (B + H_t) feet*

	(1) Rock Condition	(2) Rock Load H_p, feet	(3) Suggested Spacing of Sets, feet
(1)	Hard and intact	zero	Usually no support
(2)	Massive, moderately jointed	0 to 0.25B	If required, 5 ft.
(3)	Stratified or schistose	0 to 0.5B	If required, 5 ft.
(4)	Moderately blocky and seamy	0.25B to 0.35 (B + H_t)	Spacing 4 ft. to 5 ft.
(5)	Very blocky and seamy	0.35 (B + H_t) to 1.10 (B + H_t)	Spacing 2 ft. to 4 ft.
(6)	Unconsolidated or completely crushed	1.10 (B + H_t)	Spacing 2 ft.

*Back is assumed to be located below water table. If back is located permanently above the water table, values for (4) to (6) can be reduced by 50 percent.

(7) Squeezing ground   Exceeds $1.10(B + H_t)$ Spacing 2 ft. or less

  (a) Moderate depth $(1.10 \text{ to } 2.10)(B + H_t)$

  (b) Great depth $(2.10 \text{ to } 4.50)(B + H_t)$

(8) Wet, competent rock   May fall in classifications (1) to (5) or (7)   See (1) to (5) or (7)

(9) Swelling rock H_p up to 250 ft.   Spacing 1 1/2 ft. to 2 ft.

(10) Wet, unconsolidated or crushed materials $1.10(B + H_t)$ or more   Spacing 2 ft. or less

Table 2 -- Allowable Stresses for Air-Dried Mine Timber, lb. per sq. in.*

(1) Species**	(2) Extreme Fiber in Bending	(3) Comp. Perp. to Grain	(4) Comp. Parallel to grain	(5) Column Formula,*** P/A equals
Ash, commercial white	1100 (7000)	450 (1300)	1300 (6400)	$1300(1 - L/60d)$
Cedar, western red	800 (5300)	230 (610)	900 (5000)	$900(1 - L/60d)$
Cypress, southern	900 (7200)	170 (900)	1100 (6400)	$1100(1 - L/60d)$
Douglas fir, coast Close grained	1700 (8100)	345 (910)	1600 (7400)	$1600(1 - L/60d)$

*Modified after Wood Handbook, 1935 and Staley, 1949, Mine Plant Design.
**Values in ( ) are ultimate strength.
***L = unsupported length of column, inches.
   d = least dimension of column or diameter of round, inches.

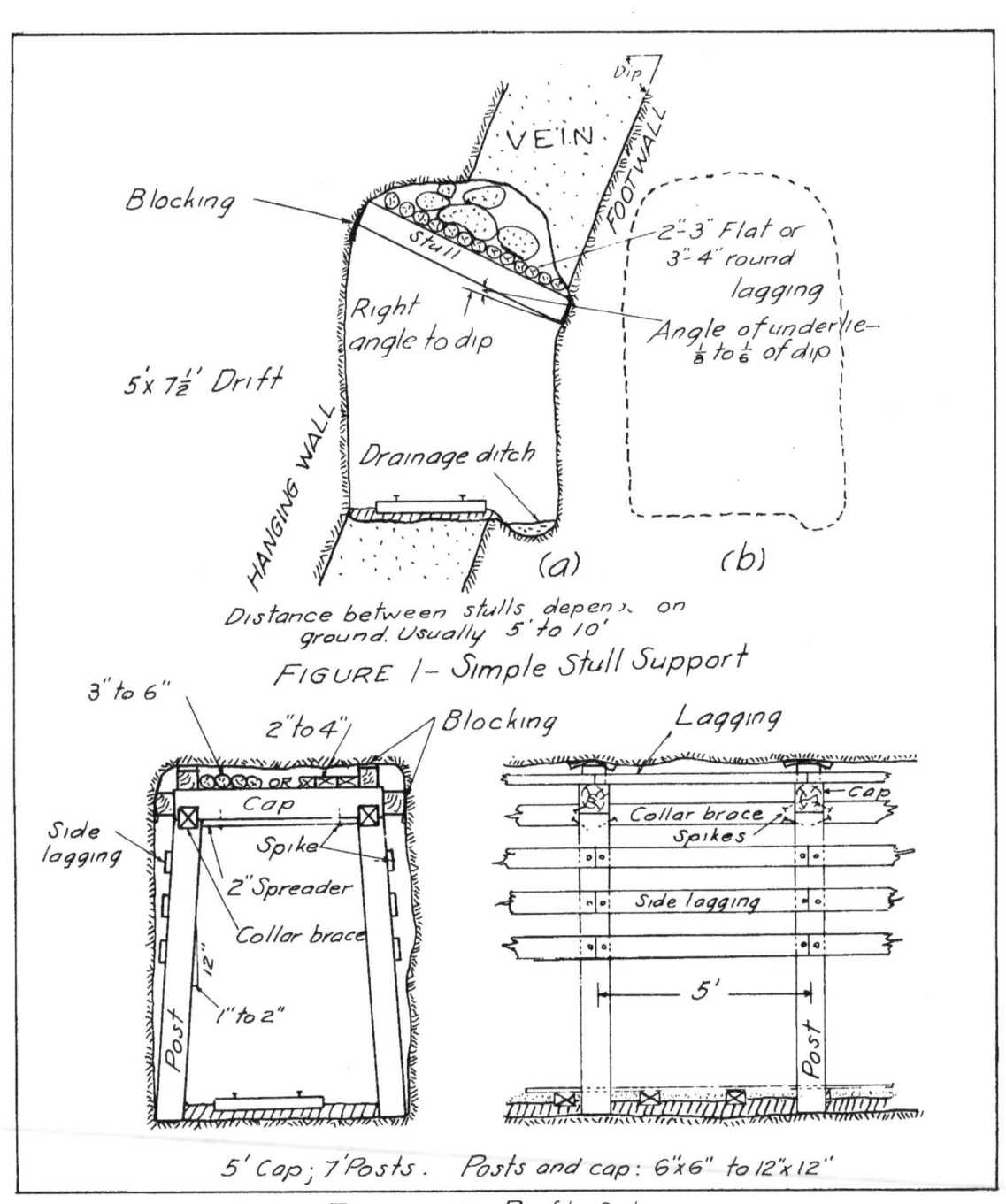

FIGURE 1 – Simple Stull Support

FIGURE 2 – Drift Set

Douglas fir, Rocky Mt.	1200 (6300)	310 (820)	1200 (6100)	1200 (1 - L/60d)
Fir, commercial white	1200 (6300)	300 (610)	900 (5400)	900 (1 - L/60d)
Hemlock, western	1100 (6800)	220 (680)	1200 (6200)	1200 (1 - L/60d)
Maple, sugar and black	1200 (8900)	450 (1500)	1400 (7200)	1400 (1 - L/60d)
Oak, commercial red, white	1100 (8100)	450 (1300)	1300 (7000)	1300 (1 - L/60d)
Pine, western white, northern white, ponderosa, sugar	800 (6000)	150 (600)	900 (5100)	900 (1 - L/60d)
Pine, Norway	900 (9400)	150 (830)	900 (7300)	900 (1 - L/60d)
Pine, southern yellow, longleaf	1300 (9300)	260 (1200)	1300 (8400)	1300 (1 - L/60d)
shortleaf	1000 (7700)	170 (1000)	1100 (7100)	1100 (1 - L/60d)
Spruce, red, white, Sitka	1000 (6700)	180 (710)	1100 (5600)	1100 (1 - L/60d)
Tamarack	900 (8000)	220 (990)	1000 (7200)	1000 (1 - L/60d)

Table 3 (after Proctor and White, 1946, p. 232) gives data for a few commonly encountered rock-types. When the material loosens and falls or exerts a swelling action through alteration, the weight per cubic foot will be less. Just what this decrease is assumed to be will depend on the judgment and experience of the user of these data.

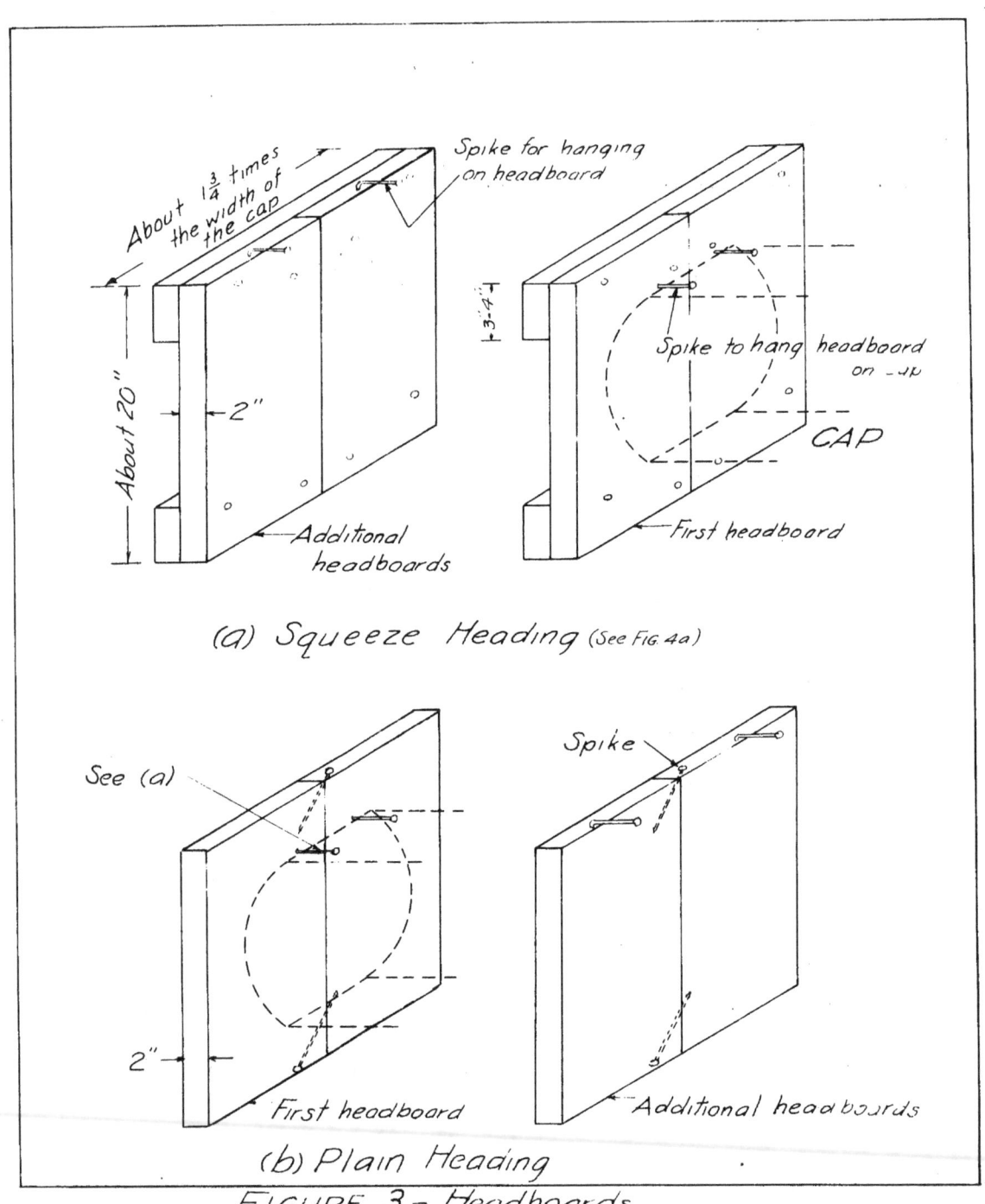

FIGURE 3 - Headboards

Table 3 -- Weights of Rocks

Material	Weight in place, lb. per cu. ft.
Basalt	175-192
Dolomite	131-168
Gneiss	165-182
Granite	145-176
Limestone	131-168
Sand, dry	120
wet	133
Sandstone	125-168
Schist	168-182
Shale	125-168
Slate	170-180

To illustrate the application of the data in Table 1 and Table 2, I have selected a timber set for the drift shown in Figure 5e. This same information will be used again for a concrete set.

Solving a problem of this kind involves the application of several simple strength of materials formulas. These are:

(1) Bending moment formulas which depend on the loads represented by W_A, W_B, or W_C.

(2) Column formula for the load on the posts resulting from W. (strictly speaking, W is the vertical component and should be resolved along the post. The resulting slight increase, however, may be neglected.)

(3) Bearing between the posts and the cap (compression perpendicular to the grain in the cap).

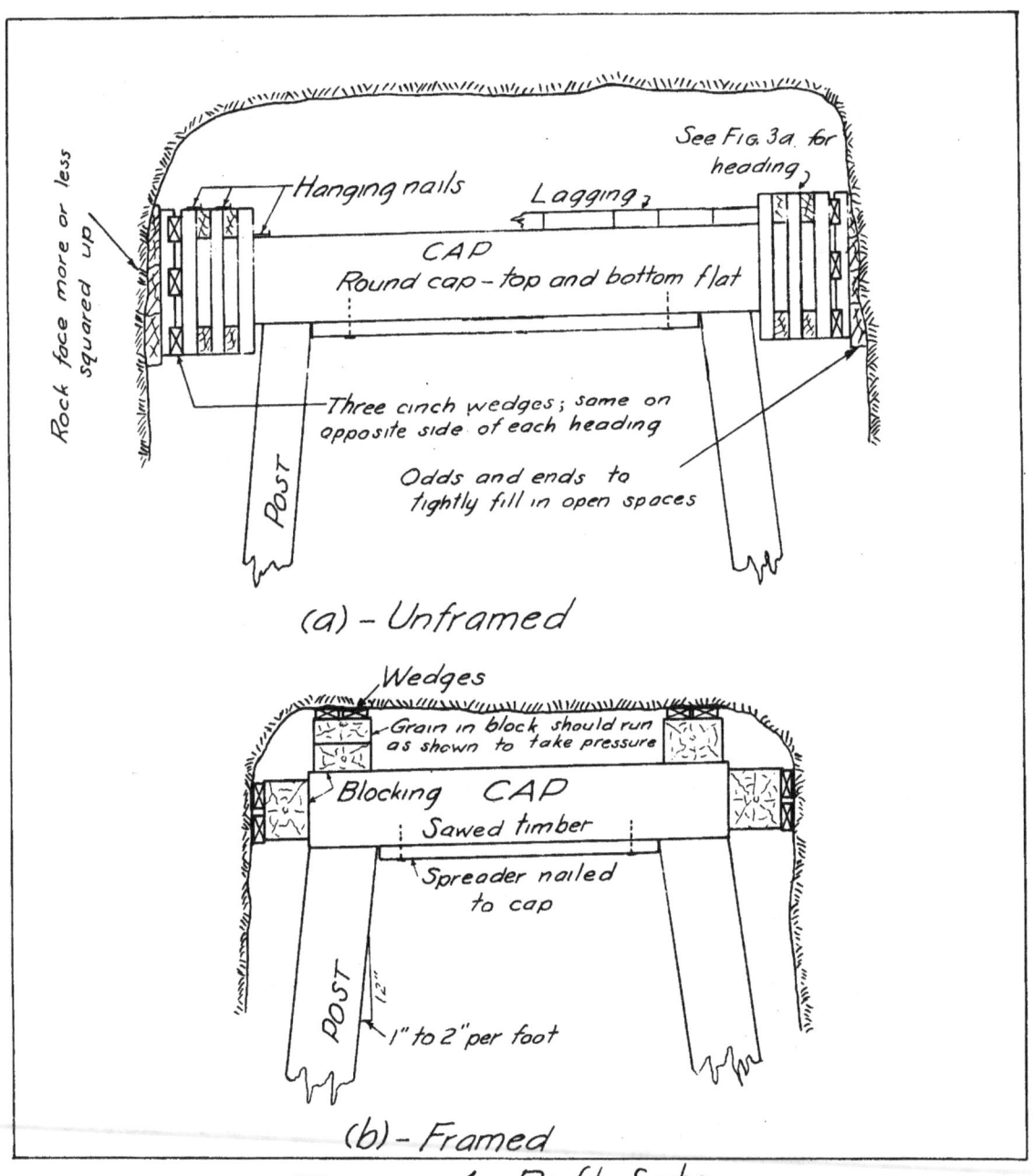

FIGURE 4 — Drift Sets

(4) Bending in posts resulting from side pressure and bearing of post against spreader (design here is influenced by compression perpendicular to grain of the post). A more refined procedure, which would combine the direct stress W with the bending in the posts, is not considered necessary here because bending is so much greater; but in some instances this procedure should be investigated.

To arrive at the size of the members making up the cap and posts, I shall take several minor liberties when applying the formulas. This deviation is justifiable on the grounds that H_p and W are not subject to exact determination and that all assumptions are on the conservative side.

SELECTION OF CAP

Known and assumed data are:

(1) Width of drift at the top = l* = 6 ft. 3 in.

(2) Average weight of fallen and compacted rock with additional squeezing effect = w = 160 lb. per cu. ft.

(3) Distance between sets, center to center = c = 5 ft.

(4) $H_p = 0.31 (B + H_t) = 0.31 (8 + 9\ 1/4) = 5$ ft. 5 in.

(5) Height of section A = h_a = 2 ft. 3 in.

(6) Height of section B = h_b = 3 ft. 2 in.

Bending moment formulas (l is in inches):

$M_A = 1/8\ W_A\ l$, in.-lb.

$M_B = 1/6\ W_B\ l$, in.-lb.

$M_C = 0.144\ W\ l$, in.-lb.

$M\ = 1/6\ s\ b\ d^2$, in.-lb.

---

*Lower case letter "l". In typescript the same symbol is used for the figure "1". Reference to the figures and the context of the formula will usually indicate which is meant.

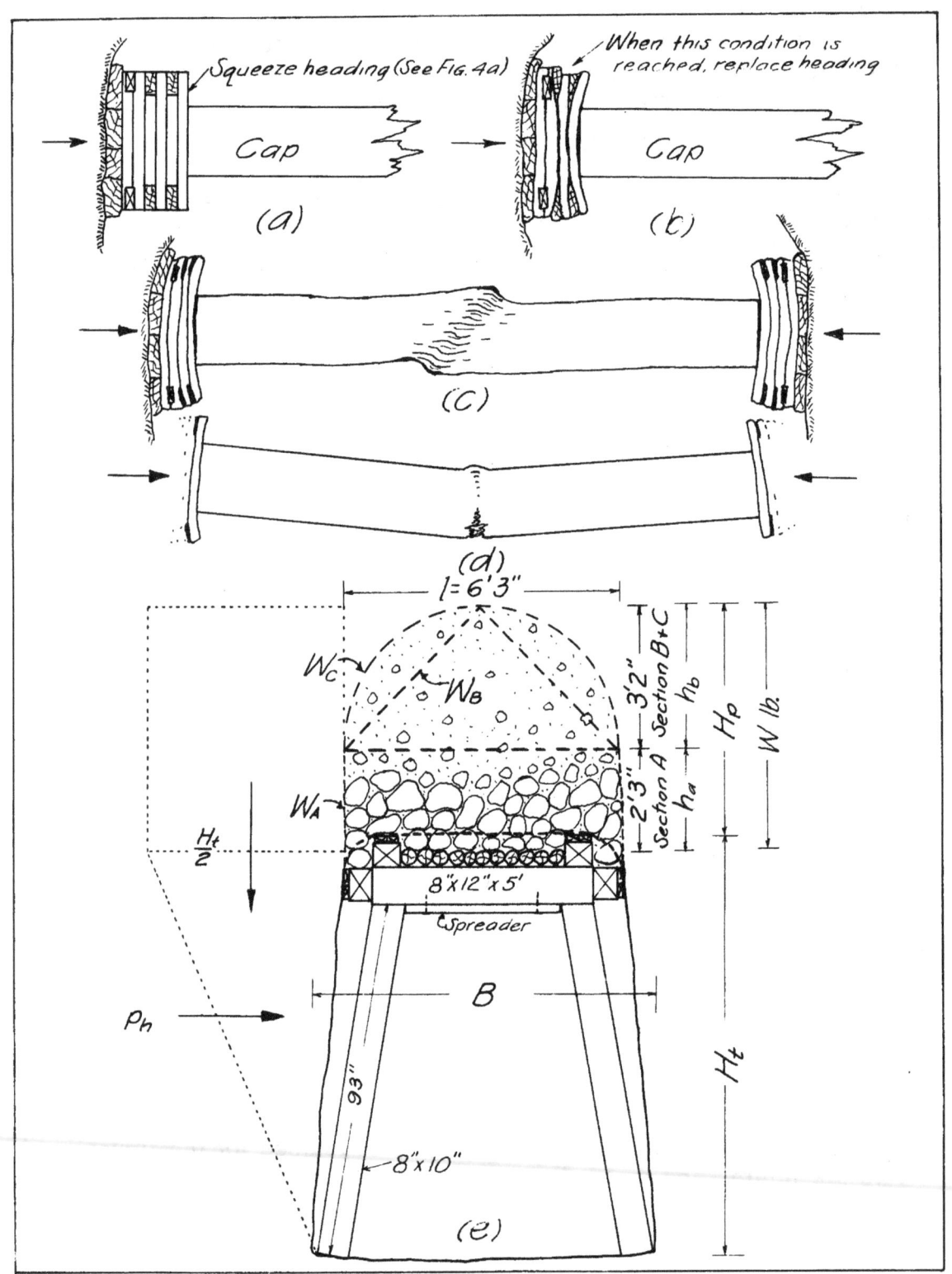

FIGURE 5 – Effect of Heavy Ground

where,

>s = Allowable working stress in extreme fiber (see Column 2, Table 2). (See later suggestion recommending higher percentage of ultimate strength for s).
>
>b = Width of cap, inches.
>
>d = Depth of cap, inches.
>
>l = Width of opening, inches.

The timber species will be Rocky Mountain Douglas fir (commonly called red fir). For this material s = 1200 lb. per sq. in.

(All calculations are made with a slide rule).

$W_A = h_a \, l \, c \, w$ = 2 1/4 x 6 1/4 x 5 x 160 = 11,240 or 11,200 lb.

$W_B = (1/2 \, l \, h_b) \, c \, w$ = 1/2 x 6 1/4 x 3 1/6 x 5 x 160 = 7,930 or 7,900 lb.

$W_C = (\frac{\pi l^2}{8}) \, c \, w$ = 1/8 x π x (6 1/4)^2 x 5 x 160 = 12,280 or 12,300 lb.

To illustrate the calculations for a cap, we shall assume that the rock eventually stabilizes with an arching effect similar to W_C. An identical procedure would be followed for W_B.

M_A = 1/8 W_A l = 1/8 x 11,200 x 75 = 105,000 in.-lb.

M_C = 0.144 W_C l = 0.144 x 12,300 x 75 = 132,800 in.-lb.

M_{total} = M_A + M_C = 105,000 + 132,800 = 237,800 in.-lb.

By applying $M = 1/6 \, s \, b \, d^2$, the cross sectional size of the cap is obtained. Because there are two unknowns in the formula, a trial and error solution must be computed. It is always desirable to take d equal to or greater than b.

When the formula is rewritten,

$$d = \sqrt{\frac{6 M}{s \, b}}, \text{ inches}$$

Assume b to be 10 inches.

(Timber for most underground uses is unsurfaced. Therefore, its dimensions are nominal or rough. If surfaced material is used, its dimensions are usually 1/2-in. less than the nominal for sizes other than planks; so for dressed lumber the surfaced dimensions must be used).

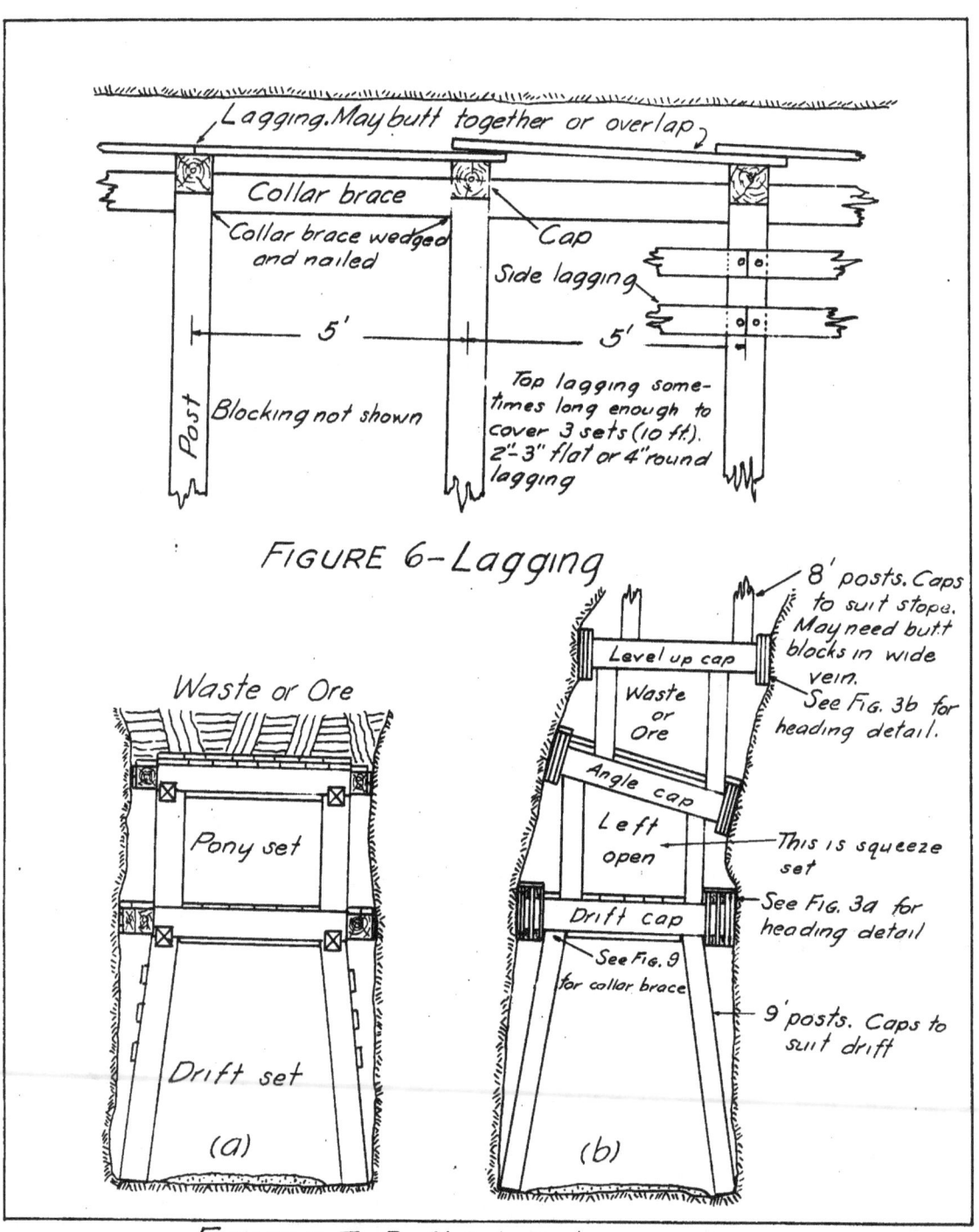

FIGURE 6 – Lagging

FIGURE 7 – Drift Sets Below Stopes

$$d = \sqrt{\frac{6 \times 237,800}{1200 \times 10}} = 10.8 \text{ in.}$$

Commercial sizes are customarily available (these dimensions change by 2-in. increments). If so, d would become 12 in. Thus it would appear that a 10- x 12-in. cap would be needed. But lumber is sold on the basis of boardfeet. A 10 x 12 represents 10 boardfeet per foot.

Eight inches for b will be tried. With this value d = 12.2 in. This figure is sufficiently close to 12 in. that an 8- x 12-in. cap would do. Here but 8 boardfeet per foot is required, thus saving the cost of two boardfeet.

As shown in Figure 5e the cap is 8 in. x 12 in. x 5 ft.

SELECTION OF POSTS

Posts are ordinarily selected to act as columns which must resist bending from the side pressure p_h, and are also checked for bearing against the spreader on the bottom side of the cap.

The total load on the cap is equally supported by the two posts, which assumes the end blocking on the caps has loosened.

$$P = \frac{W_A + W_C}{2} = \frac{11,200 + 12,300}{2} = 11,750 \text{ lb. on each post.}$$

Consulting Table 2, the column formula for Douglas fir, Rocky Mountain species is found to be,

$$P/A = 1200 (1 - L/60d), \text{ lb. per sq. in.}$$

$$= 1200 - 20L/d, \text{ lb. per sq. in.}$$

where,

P = load on post = 11,750 lb.

A = area of post, sq. in.

L = unsupported length of post, in. = 93 in. (See Fig. 5e).

d = least dimension of post, in.

b = other dimension of post, in.

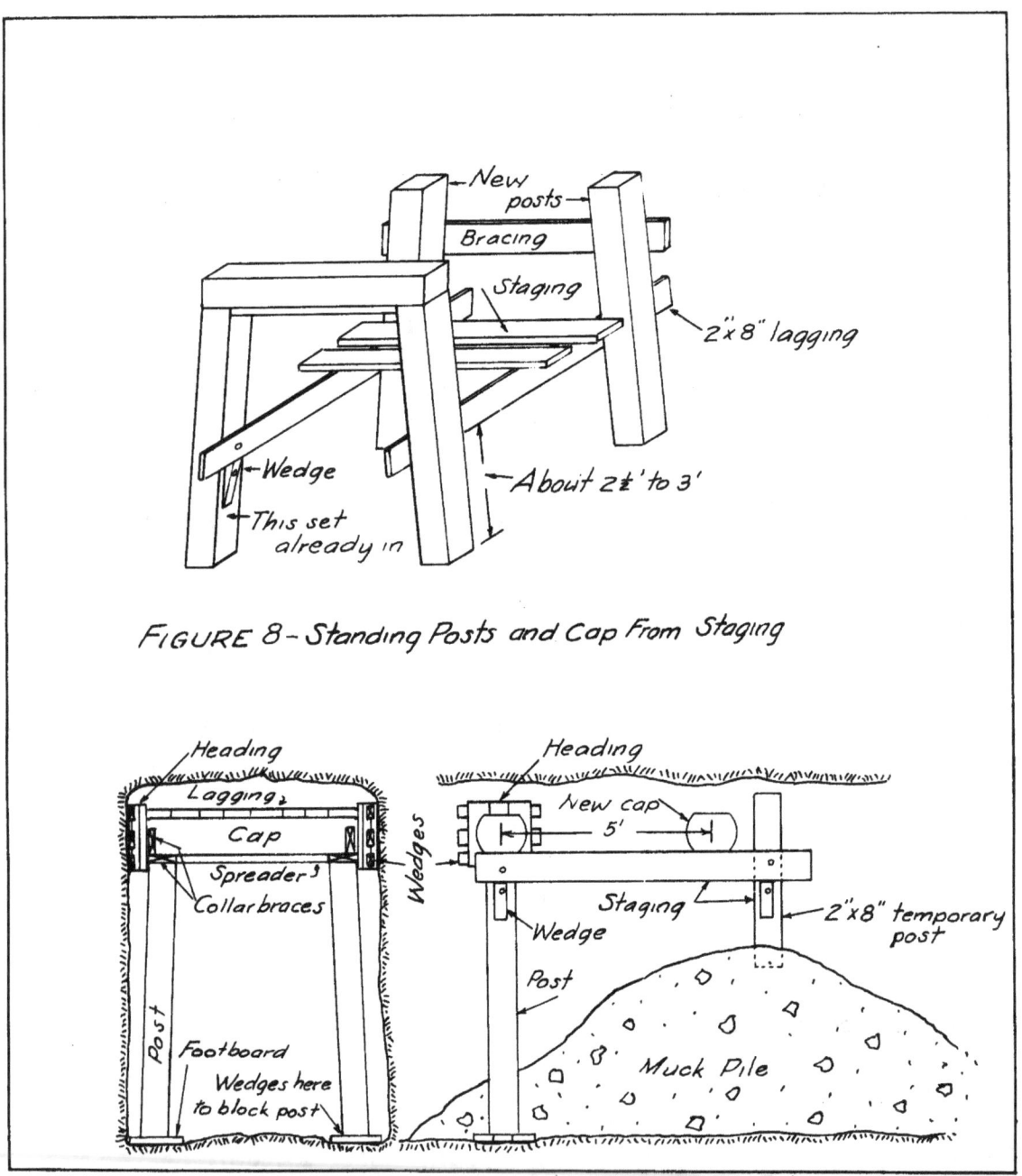

FIGURE 8 — Standing Posts and Cap From Staging

FIGURE 9 — Placing Cap From Muck Pile

Experience shows that bearing against the cap (design of joint or connection), not the member acting as a column, will usually dictate the post size. Checking for dimensions to satisfy bearing perpendicular to the grain should be the first step. From Table 2, the allowable working stress perpendicular to the grain of Douglas fir is 310 lb. per sq. in.

$$\frac{11,750}{310} = 38 \text{ sq. in.}$$

One dimension of the post must equal the width of the cap (b = 8 in).

$$\frac{38}{8} = 4.7 \text{ in.}$$

But 6 in. is the smallest mill size nearest to 4.7 in. The post is tentatively taken at 6 x 8 in.

These values are now used to check the post as a column.

$$1200 - \frac{20 \times 93}{6} = 890$$

but

$$\frac{11,750}{6 \times 8} = 244$$

$$890 > 244$$

Acting as a column, the 6 x 8 selection is more than ample. Anything less than this size will crush the cap fibers through bearing.

Post in bending

In Figure 5e, the dotted rectangular section plus the dotted triangular portion would, under certain circumstances, cause the side pressure on the post. Several approximations will be used to establish the maximum load against the post.

(1) Assume the rock alters and decomposes into a moist sand. Proctor and White (1946, p. 63) suggest that the average pressure resulting from this column of sand is represented by $p_h = 0.3 w (0.5 H_t + H_p)$, lb. per sq. ft., where w = the weight of one cubic foot of sand.

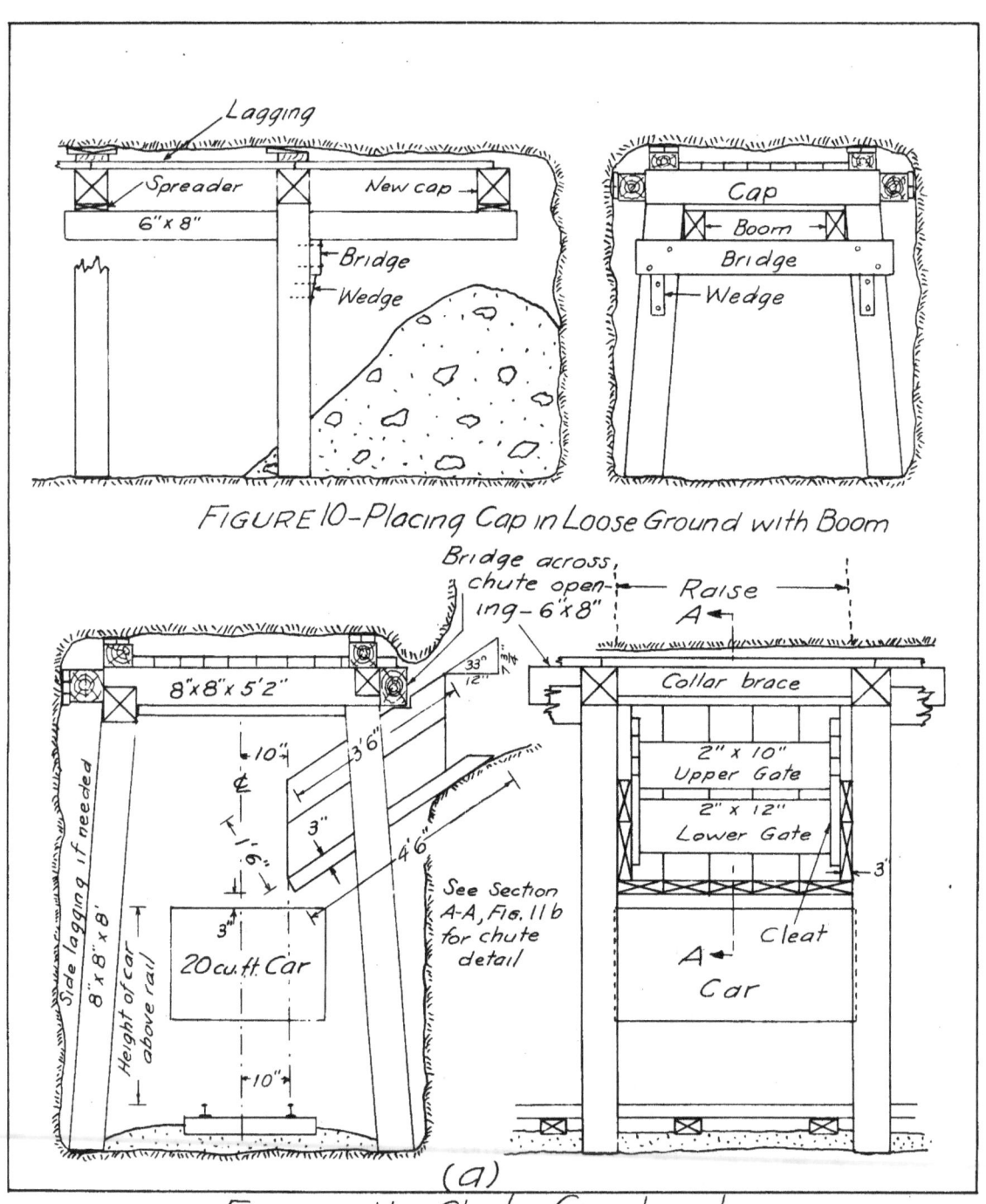

FIGURE 10 – Placing Cap in Loose Ground with Boom

FIGURE 11 – Chute Construction

(2) Moist sand in place has a bank weight of about 130 lb. per cu. ft. (Basic Estimating, (no date), p. 14). $p_h = 0.3 \times 130 \times (0.5 \times 9 + 5' 5") = 386$ lb. per sq. ft. _average pressure_. Total W = p_h c times height of post, lb. = $386 \times 5 \times 7\,3/4 =$ 14,900 lb.

(3) The effective loading on the post is similar to the dotted triangle shown in Figure 5e. The base of the triangle is taken equal to the length of the post or 93 inches. For this type of loading, where load is increasing uniformly toward one end,

$$M = \frac{2}{9\sqrt{3}} \; W\,l = 0.1283 \; W\,l, \text{ in.-lb.}$$

$$= 0.1283 \times 14,900 \times 93 = 170,800 \text{ in.-lb.}$$

$$d = \sqrt{\frac{6 \times 170,800}{1,200 \times 8}} = 10.3 \text{ or } 10 \text{ in.}$$

Therefore, an 8- x 10-in. post will be required for the conditions necessitating the 8- x 12-in. cap.

Force acting on spreader

The force acting here is represented by 2/3 W, pounds.

$$P = 2/3 \; W = 2/3 \times 14,900 = 9,933 \text{ lb.}$$

From Table 2, the allowable stress perpendicular to the grain for the material under consideration is 310 lb. per sq. in. This value is conservative.

$$\text{Area for end of spreader} = \frac{9,933}{310} = 32 \text{ sq. in.}$$

For the 8- x 12-in. cap,

$$\frac{32}{8} = 4 \text{ in.}$$

A spreader 3- or 4-in. thick would suffice.

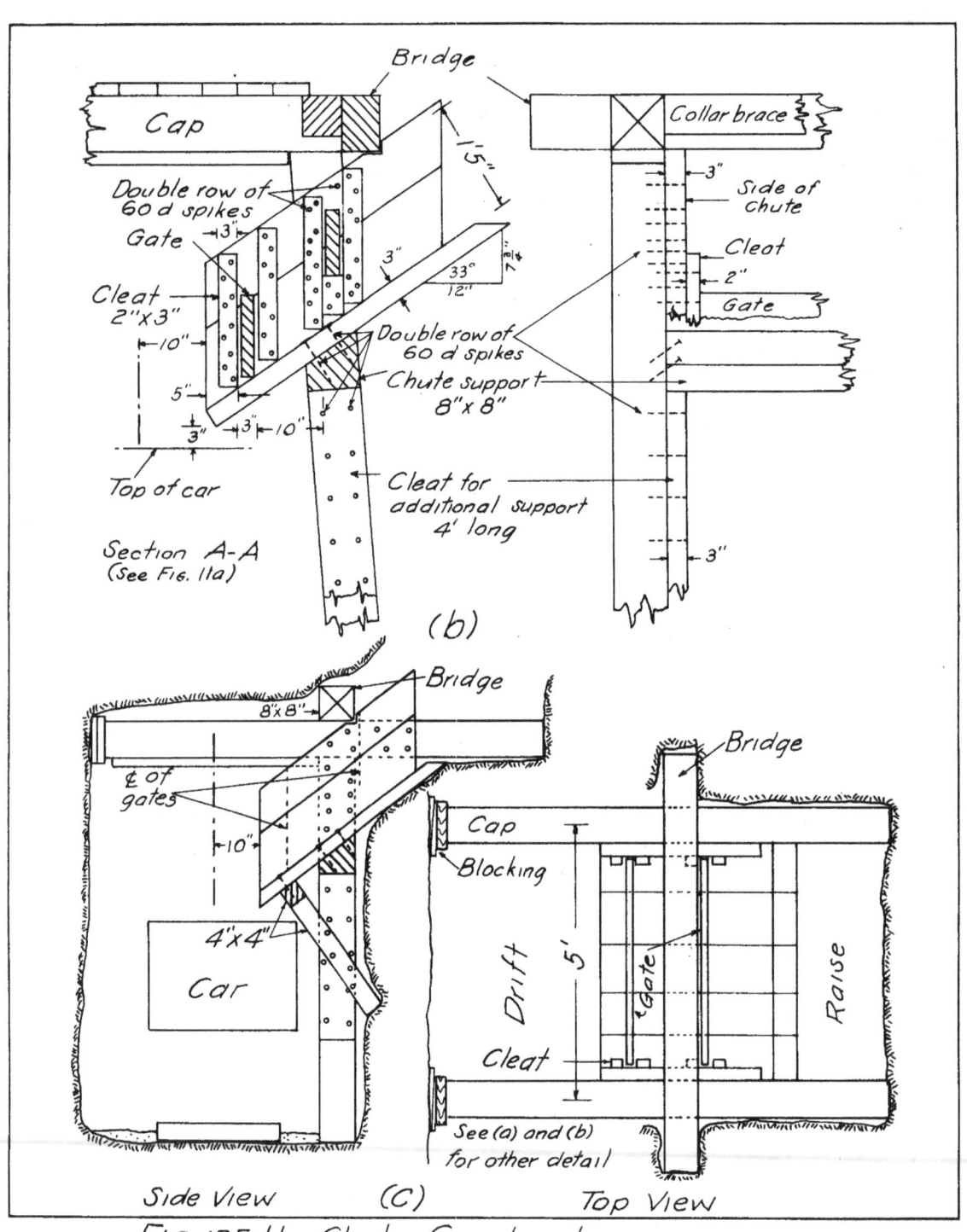

FIGURE 11 — Chute Construction

SUGGESTED CHANGE IN WORKING STRESS

The cap and posts in the preceding discussion were selected by methods using conventional working stresses for the material; these values were originally selected for use in designing various surface structures.

The allowable working stresses given in Table 2 include ample factors of safety and are used in designing structures which involve public safety and which must have a long life. These safety factors, which have been adopted after intensive investigation and observation, are for their purpose time proved.

Choice of members for underground supports should be approached with an entirely different philosophy, however. Working stresses may more closely approach the ultimate stress (see also discussion for concrete support). Because

(1) public safety is not involved;

(2) compared to surface structures, the operating life of mine supports is relatively short (as little as a few months);

(3) excepting the maximum effect resulting from rock bursts, the instantaneous and complete failure of the support with extensive closure of the opening is unlikely;

(4) timber readily indicates an increasing load (physical appearance), which provides ample time for installing additional support (installing sets between those first erected or strengthening original support);

(5) settling or disturbance of supports because of failure of the component parts, will be local in character--confined to only a few sets at the most, and after effects should not be compared to those of surface structure failures;

(6) the anticipated effective life of the opening must be carefully considered.

Data in the Wood Handbook (1935, p. 50-53) suggests that the strength of green or wet timber (depending on the species) is 50 to 60 percent of the dry strength. Usually only the prospector will use freshly cut and unseasoned timber. Larger operations provide ample time for considerable seasoning even though the timber may have been treated. In the event that green or wet material must be used, a lower percentage of the ultimate strength should probably be used. The amount of correction will depend upon the time elapsing between the support's installation and its maximum loading. During this period, the ventilating air current will produce a drying effect with a resulting increase in strength.

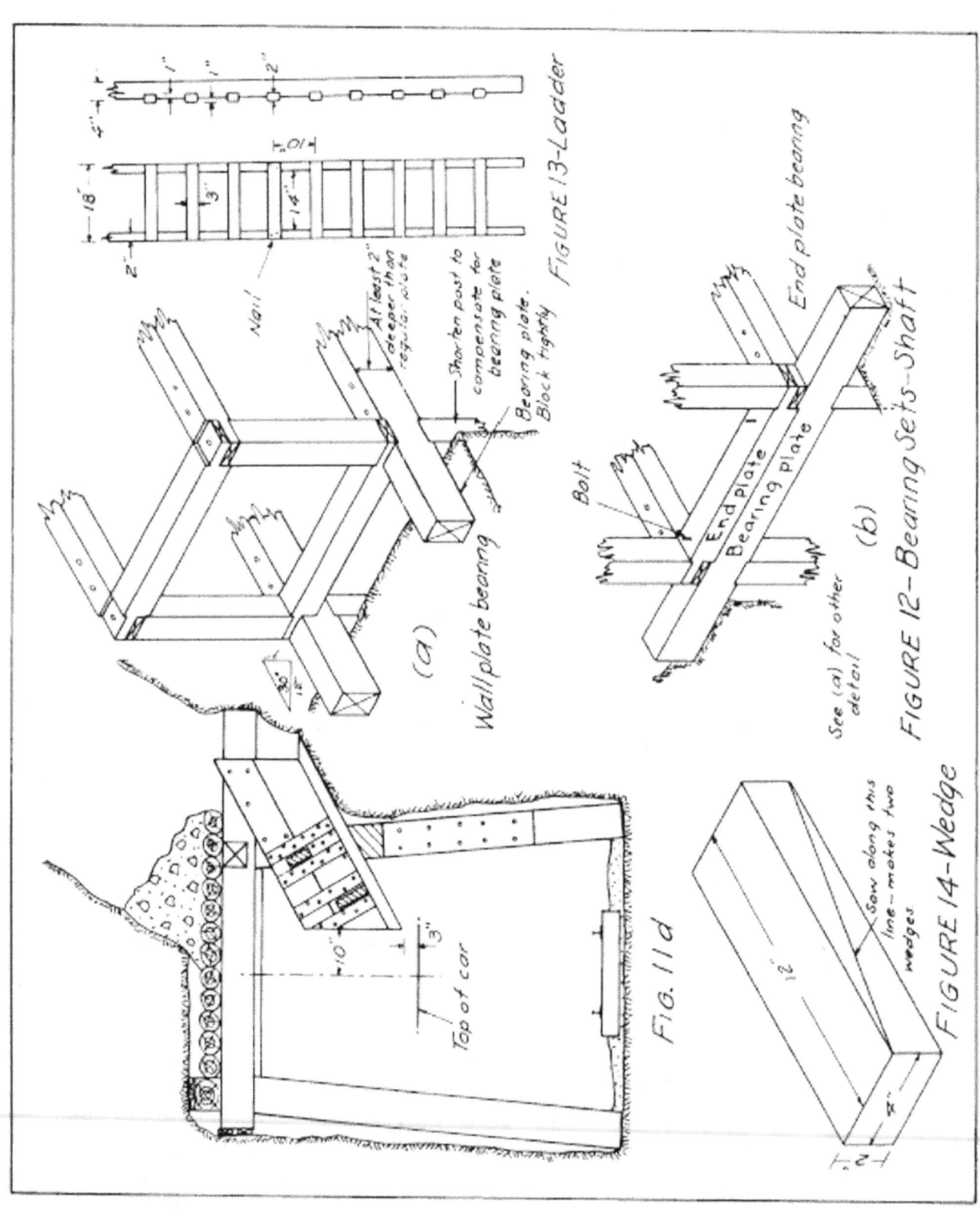

FIGURE 13 – Ladder

FIGURE 12 – Bearing Sets – Shaft

FIG. 11d

FIGURE 14 – Wedge

Under these conditions a working stress of at least 50 percent of the ultimate strength could be used, which would result in a substantial saving in timber costs.

If this recommendation were followed, the timber set previously designed would be altered to the following dimensions.

Working stress = 50 percent of 6,300 = 3,150 lb. per sq. in. = s.

Compression perpendicular to the grain = 50 percent of 820

= 410 lb. per sq. in.

Compression parallel to the grain = 50 percent of 6,100

= 3,050 or 3,000 lb. per sq. in.

Column formula, $P/A = 3,000 (1 - L/60d)$, lb. per sq. in.

Cap, 6 in. x 10 in.

Post, 6 in. x 8 in.

Spreader, 6 in. x 4 in.

If a 3-piece steel set, consisting of a horizontal cap and posts is used, it should be selected on the basis of a larger percentage of the ultimate strength. Instead of the commonly used value of s = 20,000 lb. per sq. in., at least the yield point is suggested. For the ordinary grade of structural steel, this point may be taken at 33,000 lb. per sq. in. As with the timber design, a straightline column formula is satisfactory. However, the constant k in the formula must be changed to accommodate the 33,000-lb. value used for steel (Boyd, 1917, p. 252).

When thus corrected the formula becomes,

$$P/A = 33,000 - 180 \frac{l}{r}, \text{ lb. per sq. in.}$$

where,

P = load, lb.

A = area of cross section, sq. in.

l = unsupported length of column (post), in.

r = least radius of gyration for the steel section used, in. (obtained from a steel handbook).

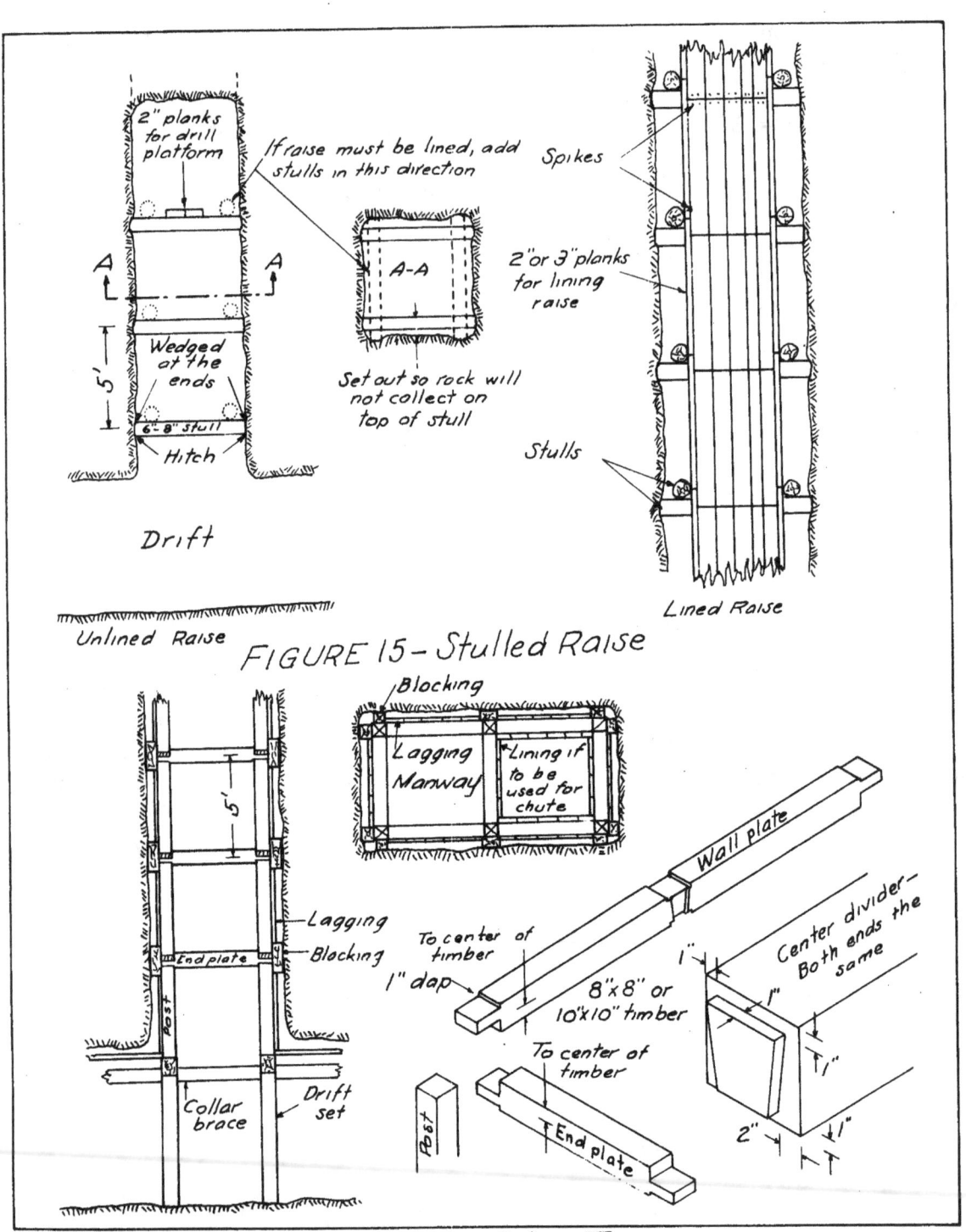

FIGURE 15 – Stulled Raise

FIGURE 16 – Raise Timbering

Later, under Concrete Sets, a parallel discussion is offered for using ultimate strength design for concrete.

For bending stresses, the convenient formula when using steel is,

$$M = \frac{sI}{c}, \text{ in.-lb.}$$

where,

M = bending moment, in.-lb. (M_A, M_B and so on as previously explained).

s = working stress in steel, lb. per sq. in. = yield point = 33,000 lb. per sq. in.

I = moment of inertia, in.^{4} (This is obtained from a steel handbook*. It is usually the value associated with the x-x axis).

c = one-half of the depth, in. (from steel handbook).

---

*American Institute of Steel Construction, Manual of Steel Construction, New York.

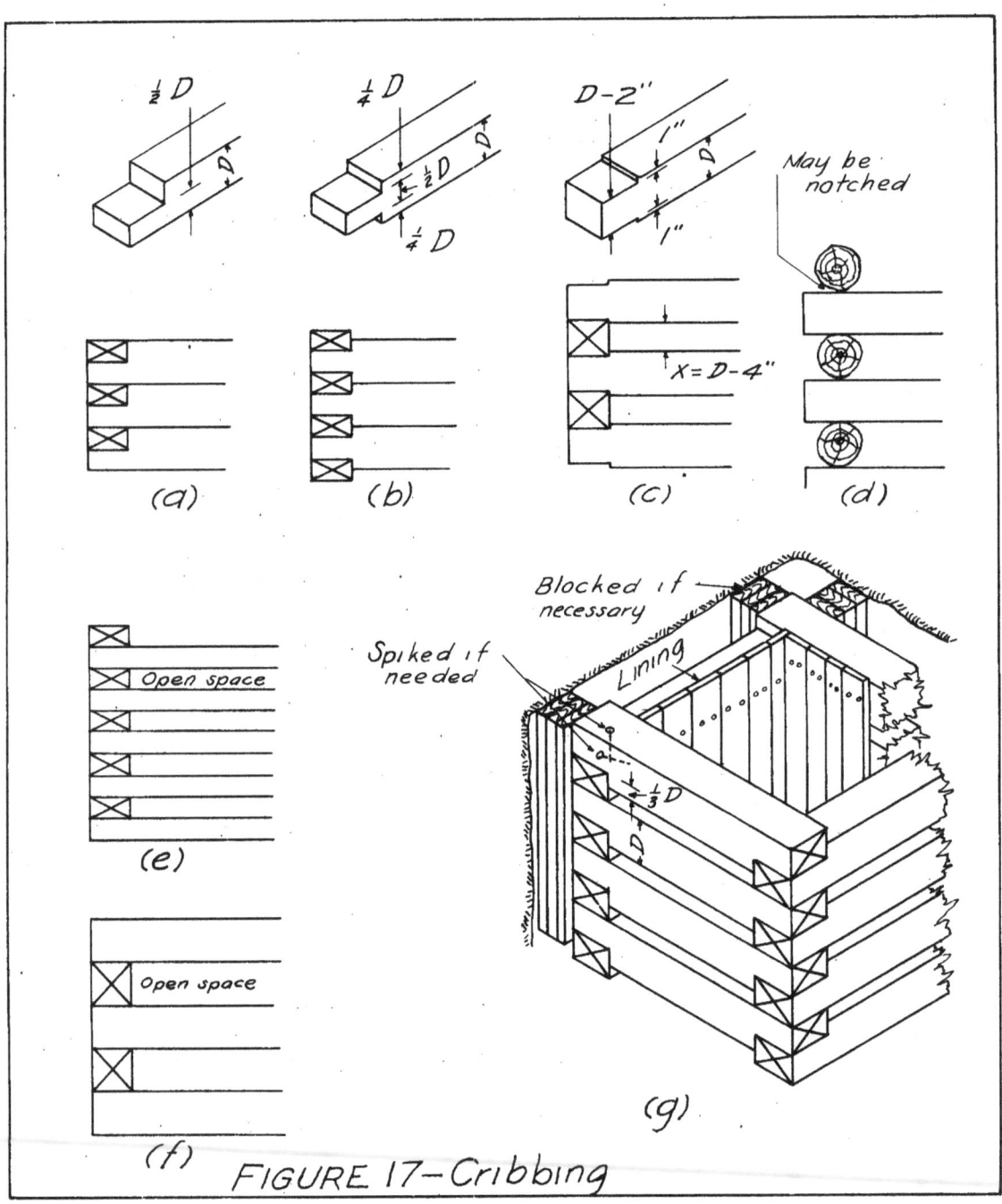

FIGURE 17—Cribbing

TREATMENT OF TIMBER

With few exceptions, humid conditions underground promote decay. Treatment of mine timber helps to slow down or prevent decay. Timber from which the bark is removed has a longer life than the unpeeled variety.

Treated timber may be obtained from local treating companies, or raw timber may be treated by the mine operator. Undoubtedly the former source is superior, for the commercial plant will usually be better equipped.

If mine timber (it should be framed before treatment) is not available from a commercial treatment plant, the following precautions should be observed if a company undertakes to treat its timber*. Also the manufacturer of timber-preserving chemicals should be consulted**.

Dipping procedure:

(1) Use a rectangular, welded steel tank 4 ft. x 4 ft. x 16 ft.

(2) Timber is best treated when either green or wet.

(3) Treat between 24-56 hours -- the longer the better.

(4) Stack treated timber so that circulation of air is retarded, to prevent rapid drying or seasoning.

The pressure-vacuum process is much superior to dipping or soaking in a tank. Ordinarily, only the well-established commercial company can afford this equipment. If the treated surface must be sawed or chopped when it is placed underground, the freshly exposed wood should be well painted with preserving solution.

Care must be used in handling and transporting treated timber to avoid excessive rupture of the treated surface.

Generally the economy of using treated timber is unquestionable, especially when the installation is to be semipermanent or permanent. In many stopes and their accessory workings (drifts, raises), long life is not necessary, however. Treatment there could be completely wasted. The life of the timber and the life of the workings must be carefully coordinated.

---

*Personal communication from Mr. Rollin Farmin, Manager of Mines, Day Mines, Wallace, Idaho.

**Osmose Salts -- Dan Kamphausen Co., Continental Oil Bldg., Denver 2, Colo.
  Wolman Salts -- Wolman Preservative Dept., Koppers Co., Inc., 771 Koppers Bldg., Pittsburgh 19, Pa.

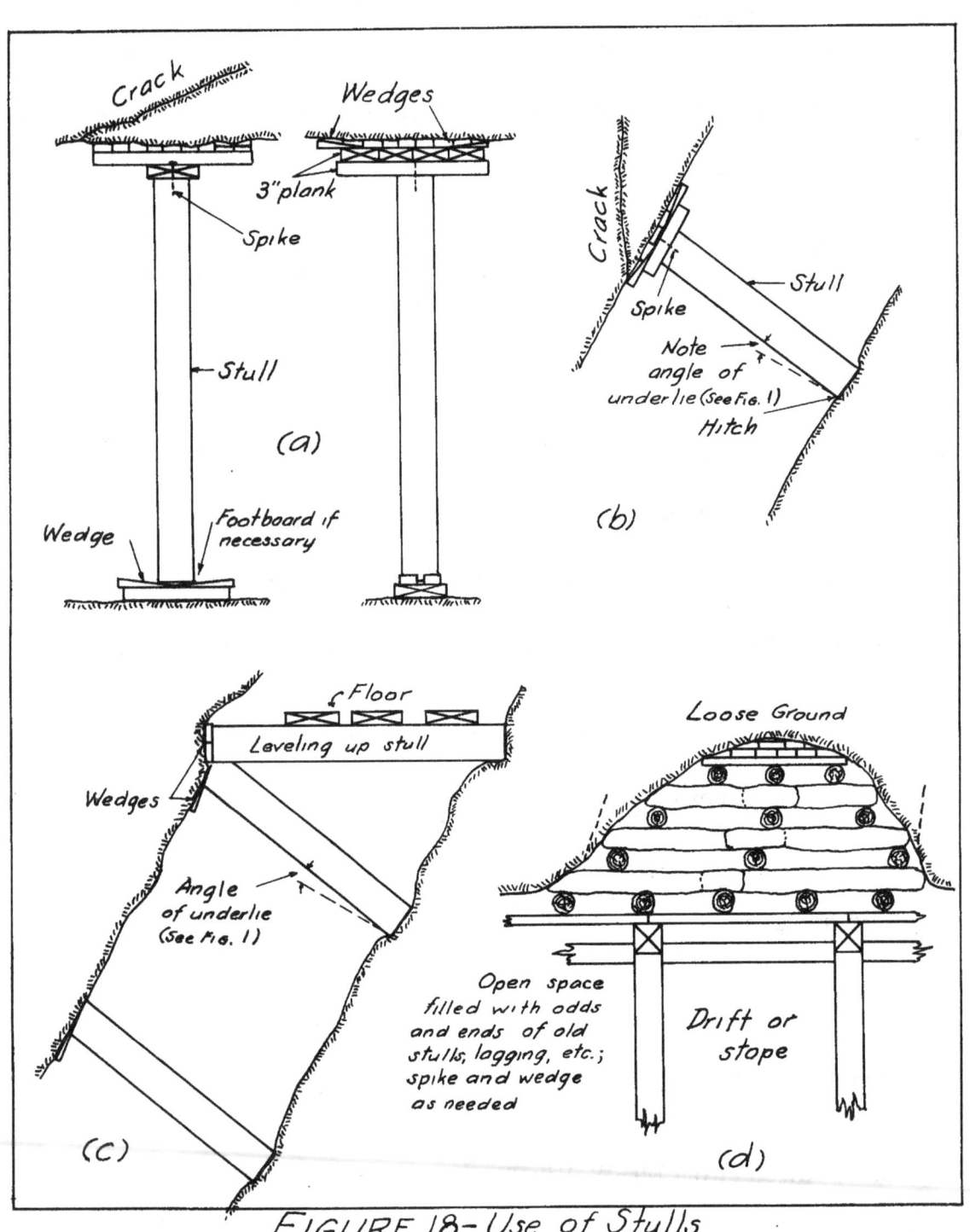

FIGURE 18 — Use of Stulls

FRAMING OF TIMBER

Nearly all mine timber requires more or less framing. Framing means cutting to size, notching, cutting daps, mortising for joints, removing slabs from round timber or any other application of a saw or ax or other work on the timber. If the drilling and blasting operation is properly planned, a minimum of subsequent underground framing will be necessary. When treated timber must be altered before installing, the freshly exposed surface should be painted with preservative.

Equipment for timber-framing is available in great variety. Machines have been designed for cutting square-set joints and other complicated work. For the small operator, and possibly even for the larger mines, hand tools are recommended. Small hand-held chainsaws, circular saws, drills, etc. are adaptable for framing. These tools may be operated by either compressed air or electricity.

A timber shop should have a swingsaw for end-cutting and a permanently mounted powersaw for making wedges (special machines are available for making wedges). Additional equipment for squaring logs and cutting planks for lagging and blocking will usually be necessary.

Before equipping a timber-framing shop, one should consult a manufacturer of sawmill equipment.

A great deal of a timberman's time underground may be saved if a plentiful supply of 2- to 4-in. thick short planks 12 to 18 in. long are cut and packaged on the surface.

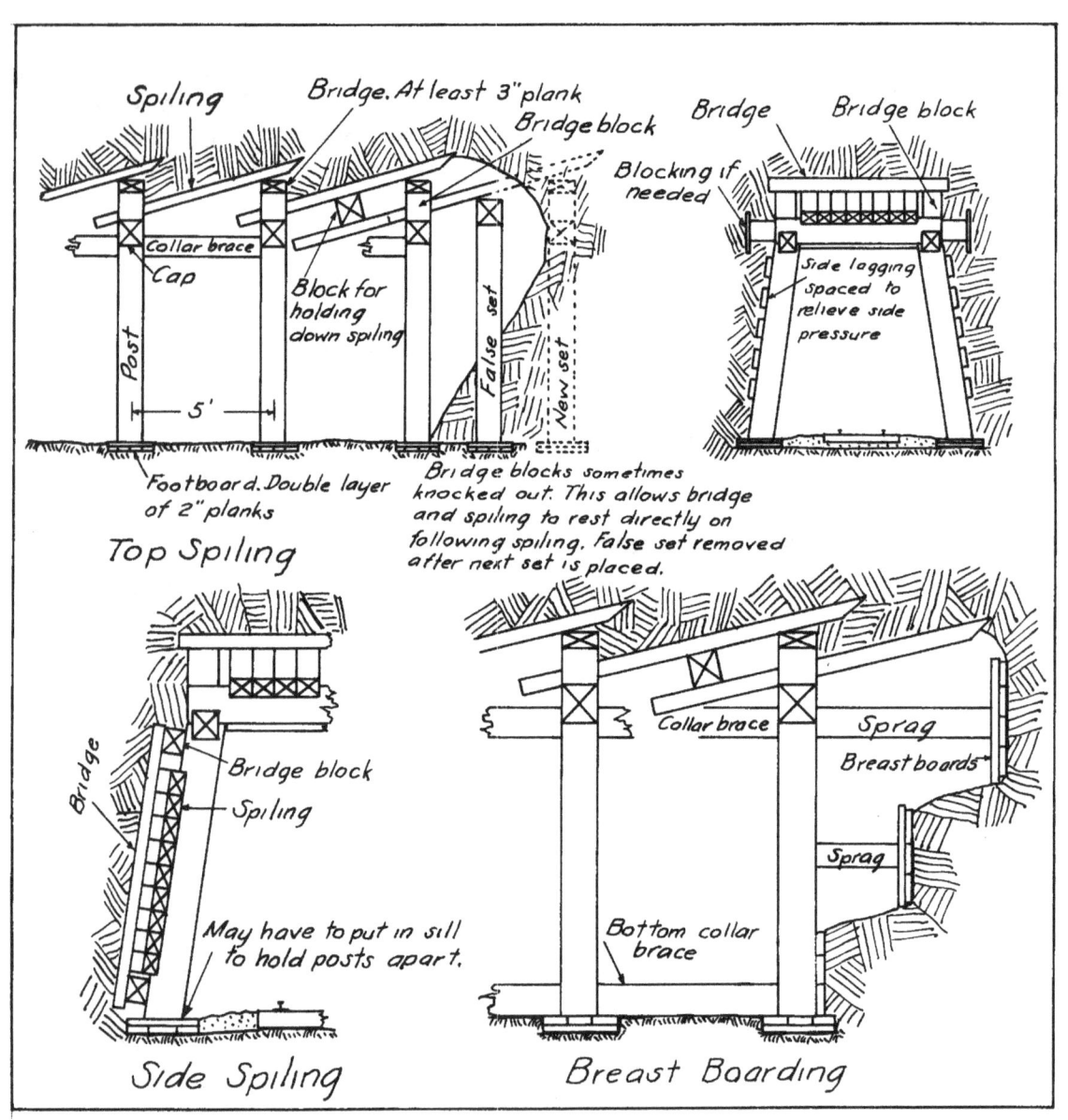

FIGURE 19 – Forepoling (Spiling)

INSTALLATION OF TIMBER

A few remarks about the erection of timber sets may be of interest to the less experienced operator.

In addition to a pick and shovel, the customary tools required by a timberman and helper are a single-bitted, 4-lb. timberman's ax and a timberman's saw (3 to 5 ft. with deep teeth). A plentiful supply of 40d and 60d spikes is also needed.

Before starting the erection of a set, whether drift, raise, stope, or other type, the various pieces of timber required should be gathered and placed conveniently at hand. Particular attention should be paid to having a plentiful supply of wedges and blocking. Interrupting operations to rustle up a wedge has resulted in the premature collapse of many an incompletely blocked set.

Figures 8, 9, and 10 offer several suggestions for erecting drift sets. If staging must be used to place the cap and top lagging, it should be securely constructed. Two men and a wet cap may well exceed 700 lb. total weight on the staging.

Blocking of sets will be treated under TIMBER REPAIRING.

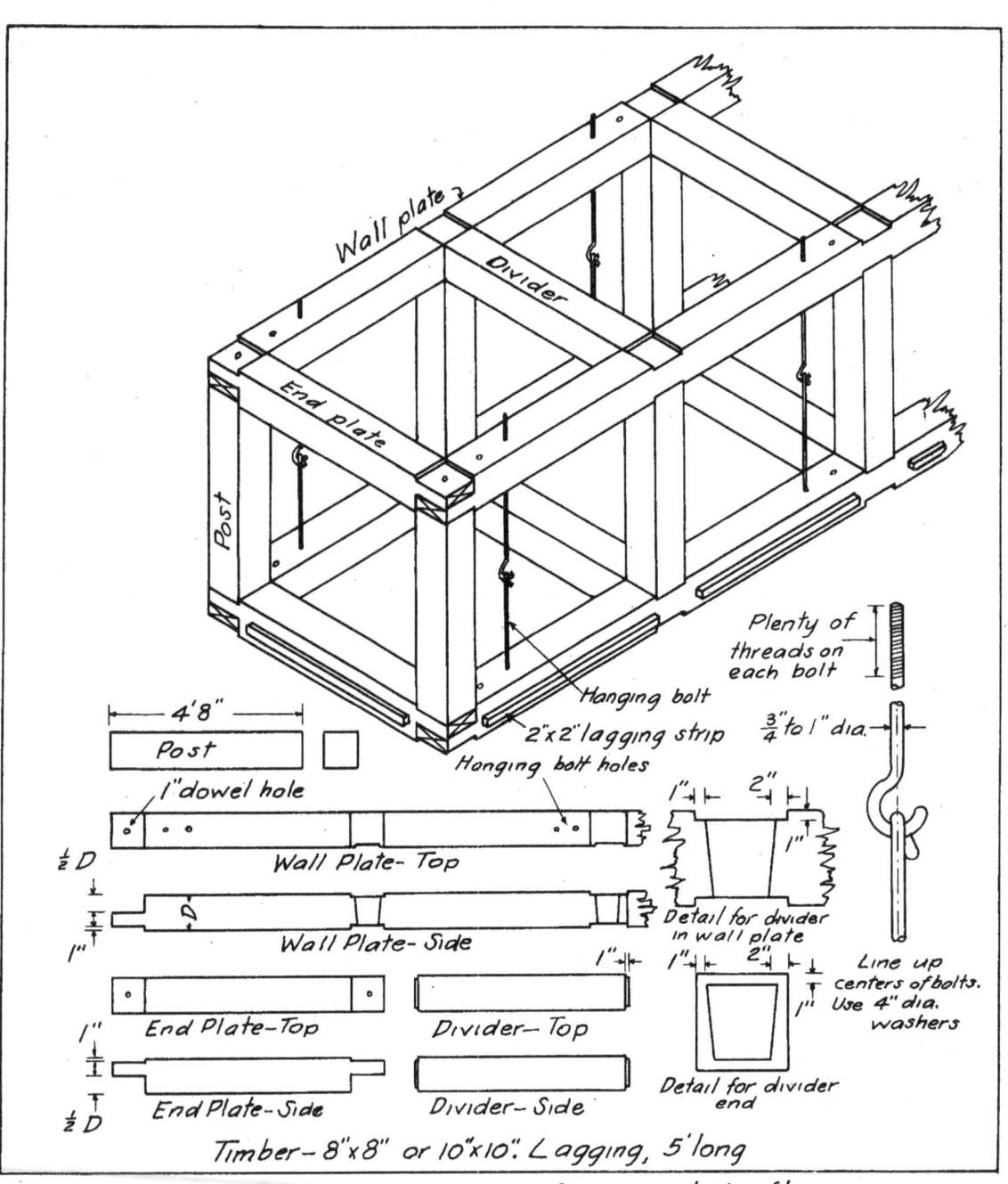

FIGURE 20—Shaft Set for Vertical Shaft

TIMBER REPAIRING

All of the timber sets to be discussed are protected against over-stressing with blocking between the set and the wall rock. Therefore, blocking serves an additional important purpose other than simply holding the member in place.

Blocking is (or should be) designed to fail before the cap, post, or other member is crushed. When the blocking has been squeezed to final deformation (see Fig. 5b and 5c), new blocking should immediately replace the old blocking before failure of the member results. In some unusual cases, failure is so rapid that replacement cannot be made in time.

Substituting fresh blocking for the deformed material is more readily accomplished before total failure takes place, and is more economical. Why wait until the cap and post, or both, must also be replaced when blocking was especially used for their protection? Not only may replacing of the set be extremely difficult, but in addition, it may be impossible to relocate the cap in its former position. Headroom and sometimes width is lost.

A "no exception" rule should be established: deformed blocking is immediately replaced before failure of the main members begins.

Enough blocking should be used so that the protected member is sufficiently cushioned to prevent its failure from the effect of ground pressure. The thickness of blocking should probably be not less than 12 in. It should consist of several pieces of blocks, planks, and wedges; this construction absorbs the pressure best. An exception to this structure is blocking used primarily to hold the set in place instead of to resist ground movement. As noted in Figures 2 and 4b, the blocking should never be placed so that the grain of the blocking is parallel to the force. Timber is 4 to 5 times stronger when force is exerted parallel to the grain than when it is exerted perpendicular to the grain.

Blocking between the top of a cap and the rock surface will have the grain in the blocking running similar to that in the cap. Here the area of the blocking could be somewhat reduced if it is to fail before squeezing the cap or causing the post to penetrate the bottom of the cap.

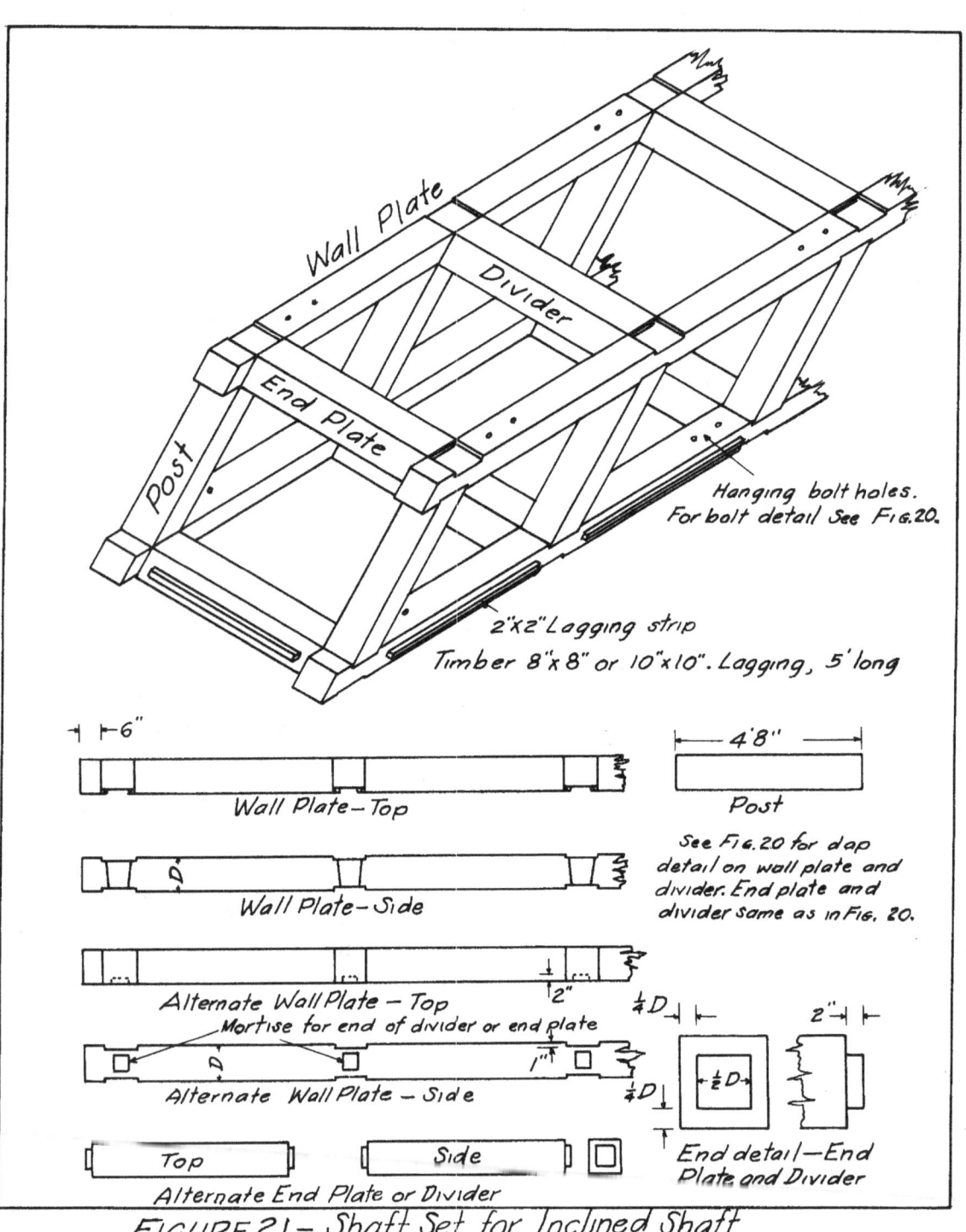

FIGURE 21 – Shaft Set for Inclined Shaft

DISCUSSION OF ILLUSTRATIONS

Many of the figures require little explanation other than that included on the drawing. For several of the illustrations a few additional remarks are appropriate.

SIMPLE DRIFT SUPPORTS (FIGURE 1)

Part (b) of Figure 1 has three advantages as an alternative location for the drift: (1) support of the back may not then be necessary; (2) when mining starts, the need for floor and back pillars will be eliminated and ore will not be tied up; and (3) stoping will interfere less with other operations through the drift. The location at (b) has one drawback for the prospector: his drift is not in the vein (Staley, 1961, p. 47). Additional crosscutting to the vein is therefore necessary at intervals, which means additional expense and loss of time, items of great importance to the prospector and miner with limited means.

Note that the angle of underlie requires a stull of greater length than the perpendicular distance between the hanging wall and the footwall. Thus, the tendency of the stull is to tighten as the load increases.

HEADBOARDS (FIGURES 3, 4a, 5a, to 5d, and 7b)

Two types of headboards for stulls or caps, shown in Figure 3, have been widely applied in many of the mines of the Coeur d'Alene district in northern Idaho. Figure 3a is generally used only on drift caps and 3b on the stope-set cap.

A squeeze heading, shown in 3a, has a definite purpose. As the wall pressure increases, the headboards making up the box-like structure gradually collapse to form a solid mass of crushed timber (Figure 5b and 5c). When the condition shown in 5b is reached—all members of the heading into tight contact with each other—the deformed headings (both ends of the cap) should be removed and replaced with new headings and blocking. If these are not replaced, the cap will ultimately be destroyed. Failure of the cap usually occurs as shown in 5c; but it may occur as in 5d. Breaking of the cap as shown in 5d may also result from a too heavy load on top of the cap (to prevent this failure, relieve the pressure by removing broken rock from above the top lagging).

The inventor of the squeeze heading is unknown. Its purpose and the thought back of it deserve much credit. But for some inexplicable reason its advantage is very seldom utilized. Instead, caps are allowed to fail and the terrific job of installing a new cap is undertaken. It would be much easier and much less costly to replace the headings as soon as the situation indicated in Figure 5b is reached.

In practice, drift caps used with squeeze headings approach 15-16 inches in diameter and about 8-10 feet in length.

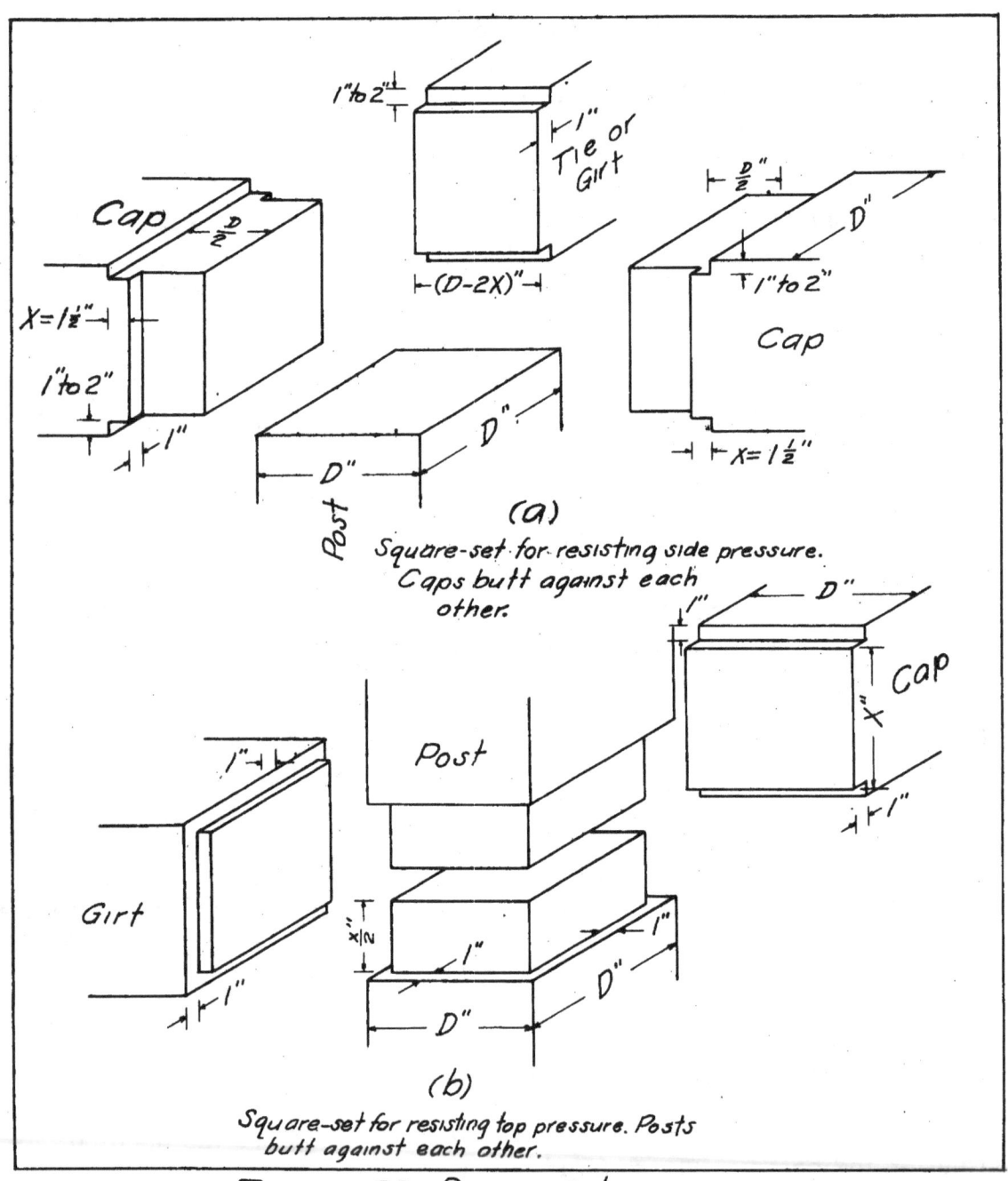

FIGURE 22—Square-Sets

Caps in the stopes (a method incorrectly classified as square-set mining, but more correctly known as stull and fill) do not need squeeze headings; plain headings as shown in Figure 3b are sufficient. Depending on the thickness of the vein, these caps may approach 16 ft. in length; their diameters are 16-18 inches. Occasionally, very short pieces of stulls (butt-blocks) are used beyond the ends of the cap to reach extra wide walls or to reduce the amount of heading and blocking.

Timber sets shown in Figures 4a and 7b are semiframed. The caps are peeled logs with squared ends and a slab removed from top and bottom (or sawed timber may be used). For drift sets, posts about 8-in. or material of larger diameter is used. In the stopes, material of about the same diameter, sawed lengthwise into two halves, is sufficient. If heavy top pressure is anticipated, posts of larger diameter are required.

SQUEEZE SETS (FIGURE 7)

Illustration 7a shows a completely framed assembly. The purpose of the pony set is twofold: (1) it provides a safe walkway and operating space for the chute puller to load cars; and (2) it provides a squeeze set to protect the main drift set if the overhead load becomes heavy. In the event this latter condition is expected, the size of the pony-set timbers should be one standard size smaller than those of the main drift set.

Figure 7b shows the squeeze set when unframed timbering is used. The purpose of squeeze sets is to insure operating access to the drift. Therefore, when the squeeze sets have collapsed to the point where destruction of the drift set is imminent, the debris resulting from the destruction of the squeeze set should be removed and a new set installed.

ERECTION OF DRIFT SETS (FIGURES 8, 9, AND 10)

Note the use of wedges to help support the staging as a safety precaution. Two heavy men and a heavy cap might cause failure of the staging. Several of the Figures illustrate other applications of wedges.

CHUTE CONSTRUCTION (FIGURE 11) (FIGURE 26m)

Figure 11 offers suggestions for selecting and designing chute detail. Chutes (c) and (d) are used in drifts where ground-support timbering is unnecessary. Usually it is possible to support the main weight of the broken ore column on solid rock rather than on the wooden chute bottom. This procedure should be considered when designing a chute. Later salvaging of usable parts should also be considered in chute design. Figure 26m offers a method for attaching the wooden chute members to a concrete post. Instead of the bolting shown there, one of the available adhesive materials may be used for attaching the wood to the concrete.

SHAFT BEARING SETS (FIGURE 12)

Shaft bearing sets are recommended at about 100-ft. intervals. They are used to support regular sets through the medium of the hanging bolts in the event that the blocking becomes loose. Either end-plate bearing or wall-plate bearing may be used.

CRIBBING (FIGURE 17)

In heavy or squeezing ground the blocking shown in Figure 17g may be replaced by or supplemented with cedar lagging, 6-in. to 8-in. diameter by 5-ft. long, split to give 4 to 6 triangular-shaped pieces. This lagging is firmly packed between the cribbing and the rock walls. Ground movement is taken up by compressing the cedar.

APPLICATION OF STULLS (FIGURE 18)

Figure 18d suggests a means of catching up a caved area following the failure of the supporting timber. Situations of this sort are best prevented at the start. Usually only about one car load of rock drops out at first. This rock is cleaned up but nothing is immediately done about catching up the back. Eventually a large cave develops which is difficult to fill, and subsequent caving of the back is hard to stop. Immediate corrective action at the start usually would have prevented further enlargement of the caved area. The extension of caving ground or the displacement of timber support may be greatly reduced if corrective measures are taken promptly.

LOOSE OR SQUEEZING GROUND (FIGURE 19)

Fault zones, squeezing ground, and similar conditions are frequently impossible to penetrate and hold by ordinary supporting procedure. This figure explains and shows how spiling may be used. Spiling is driven with a maul or sledge, or sometimes by a heavy block swung from a rope. To protect the end of the spiling, a temporary iron shoe is often provided to cap the timber during driving. When the job is completed, projecting ends are sawed off.

Shafts have been sunk through loose ground by first freezing the site. For a description of the technique, Peele (3rd Ed., 1941, sec. 8, p. 20) may be consulted. A similar procedure could be used for driving drifts through loose, wet ground.

Recently (Pynnonen and Look, 1958; Eng. and Min. Jour., Nov. 1958, p. 126; Sun and Purcell, 1959, p. 1), a method known as chemical injection has been devised for treating and stabilizing loose ground. This process appears to have considerable merit.

VERTICAL AND INCLINED SHAFTS (FIGURES 20 AND 21)

Little need be added to the information given with the drawings so far as the illustrations are concerned. Ground pressure for the first few hundred feet of a shaft is seldom important. Sets--either timber, steel, or concrete rings--are installed to divide the cross section into compartments; to provide a means of installing cage or skip guides; and to prevent rock from loosening and falling into the shaft. The spacing between sets is usually 5 ft. center to center.

Dimensions of the members will depend on the ground condition. For relatively shallow mines, dimensions about 8- x 8-in. would be satisfactory; with increasing ground pressure, 12- x 12-in. or larger may be required. Under exceptionally bad conditions, a "squeeze set" may have to be installed between the main set and the rock surface. A squeeze set consists of wall plates and end plates separated from the main set by blocking which will collapse before excessive damage results to the main set.

The practice of salvaging the hanging bolts after timbering is completed is of doubtful merit. If the blocking loosens, sets may drop out with serious consequences.

Figures 20 and 21 illustrate the design for a small tonnage operation. If the shaft is sunk with only two compartments, one would be used for the cage and car. A manway would be placed in the remaining compartment. Ladders in the manway should not exceed 20 ft. in length, be inclined, and each section start from the opposite end of a platform (Mining Laws, 1959, par. 47-405, p. 39). For economical hoisting a counterweight should be installed in the manway compartment (Staley, 1949, p. 305). Contrary to general belief this installation is neither difficult nor costly.

If a three-compartment shaft is sunk, two compartments will be available for hoisting and the third for the manway. For less than 300-500 tons per day, an operation in partial balance should be used (this is loosely spoken of as balanced hoisting). Or, if a greater tonnage is feasible, one of two arrangements is commonly used: (1) a cage with a skip suspended beneath it, or (2) a skip-cage changing arrangement. At least one cage should be continually available. For a three-compartment shaft the first arrangement is probably the better.

For the inclined shaft the arrangement of the ladders will depend on the inclination of the shaft. Up to about 45 degrees, platform interruptions are not too necessary. At steeper inclinations there should be frequent interruptions in a continuous ladder.

Hoisting through an inclined shaft is accomplished by having the skip or mancar mounted on wheels running on steel rails.

For the design of safety devices or dogs see Staley (1949, p. 250).

SQUARE-SET TIMBERING (FIGURE 22)

As indicated in Figure 22, square-sets are designed for either post loading or cap loading. If the ground load is acting at an angle so that components of the load exert heavy pressure on both the cap and the post, then a more complex set may be required. A more advanced discussion on this subject should be consulted (Peele, 1941, sec. 10, p. 213-218; Gardner and Vanderburg, 1933).

When selecting a square-set design, one should adopt the design having the fewest number of sawcuts but still providing the maximum support. An excessive number of cuts adds little to the strength of the set, and makes erection difficult. Furthermore, it is difficult to get even contact and bearing between the many faces. (In fact, too many cuts may actually prevent maximum support).

It is important to remember that only three pieces of the set are used to describe a square-set or express its efficiency or economy. For example, the boardfeet of lumber in a set refers only to one each of a cap, post, and girt. Similarly, the number of sawcuts means the total number of cuts for these same three members.

Square-sets must be tightly blocked and closely filled with hydraulic fill or waste rock. Failure to adhere strictly to this principle may result in the loss of a stope.

APPLICATION OF ROCK BOLTS (FIGURES 23 AND 24)

Figure 23 illustrates four commonly used types of rock bolts. Types (a) and (b) are most widely used*. There are many varieties of (b).

In (c) the wooden bolt shown was developed for use at the Day Mines in the Coeur d'Alene district, Idaho (Farmin and Sparks, 1953, p. 922). This bolt proved satisfactory in holding soft, wet ground in and near fault zones.

Figure 23d shows a type of bolt** which is receiving considerable attention.

Use of the PERFO bolt may be briefly described as follows: the two perforated halves are filled with sand-cement-water mortar, the halves tightly wired together, and the filled tube pushed into the drill hole. Pushing the bolt or reinforcing rod into the tube forces the mortar out through the perforations to tightly fill the drill hole and cracks. Advantages of this method are:

---

*Ohio Brass Co., Mansfield, Ohio. "Haulage Ways", Sept. 1956.
Bethlehem Steel Co., Bethlehem, Pa. "Bethlehem Mine Roof and Rock Bolts".
The Colorado Fuel and Iron Corp., Denver, Colo. "Mine Rock Bolts".
Literature available from these companies and also PERFO contains many excellent suggestions on installing rock bolts. Copies are available on request.

** PERFO Division, Sika Chemical Corp., 35 Gregory Ave., Passaic, New Jersey. "PERFO Method for Roof-Bolting".

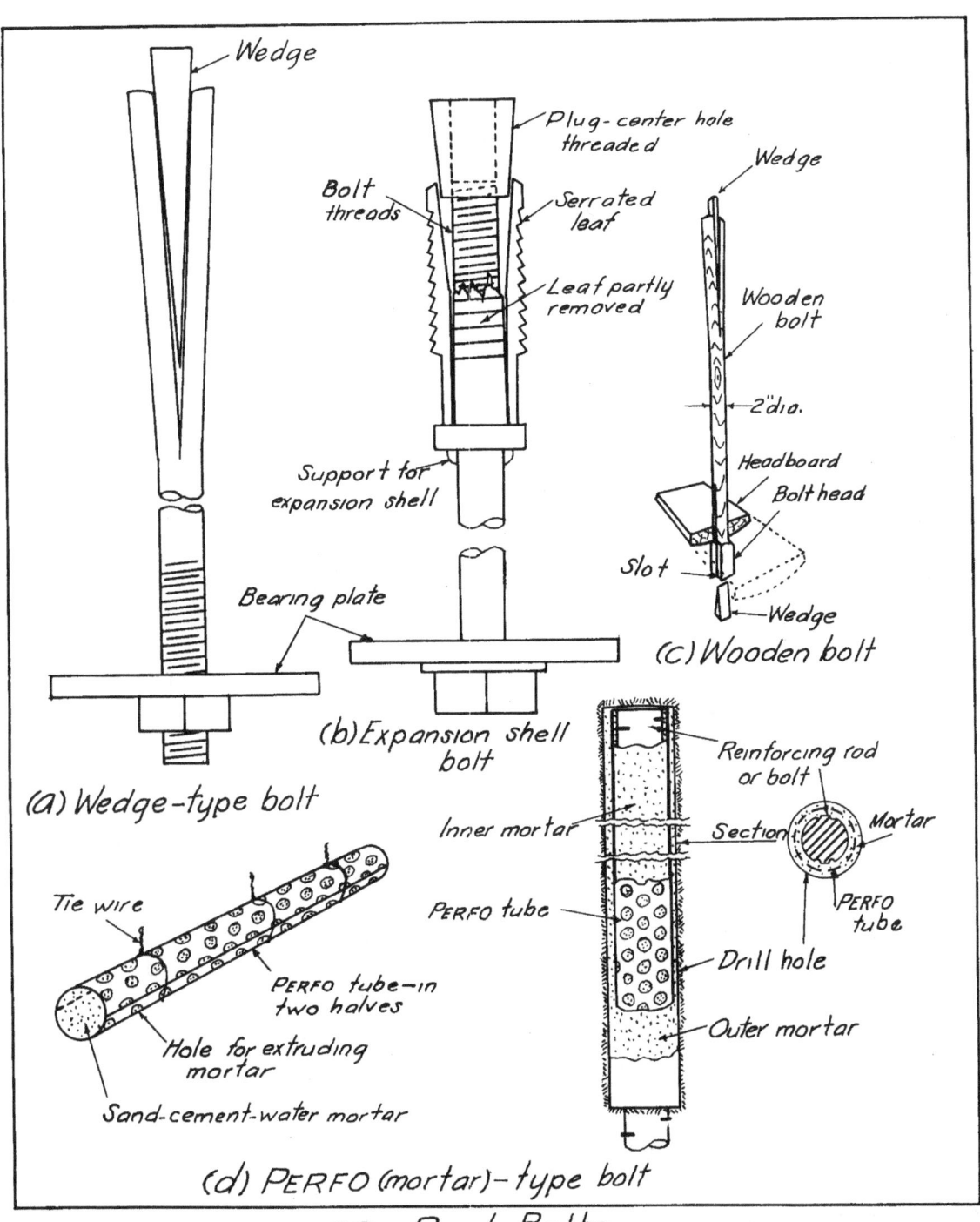
FIGURE 23 - Rock Bolts

(1) The entire length of the device supports the ground.

(2) Circulation of moist air through the hole is eliminated. (Remember the deleterious effect of moist air on many rock-forming minerals).

(3) It can be applied in soft rock.

(4) Less attention need be paid to uniformity of the diameter of the drill hole.

Unfortunately, a great deal of the more practical literature on rock bolting occurs in current technical journals or government publications of limited distribution. For this reason some parts of the article by Humphrey (May 1956, p. 491-495) will be quoted extensively.

Bolting patterns

Bolting patterns vary with rock conditions. A typical pattern in a 16-ft. opening in a coal mine would be four bolts per row, spaced 4 to 5 ft. in direction of the advance. The two center bolts would be driven vertically, or normal to the plane of the structure, while the outside bolts might be angled out about 30° to obtain anchorage over the rib. A steel angle washer is used to provide a bearing surface normal to the angled bolts.

In unstratified ground a pattern might consist of bolts driven in rows every 4 or 5 ft., one bolt vertically in the back; two bolts on each side, 4 ft. from the center bolt, at a 15° angle; two additional bolts 4 ft. down the side at a 30° angle. Special cast iron or steel beveled washers are used in that type of bolting to correct for angularity of the bolt, and to bring the bearing plate parallel to the rock's surface. (Frequently the bolts are all driven normal to the surface of the opening).

In larger openings, such as motor rooms, intersections, and shaft stations, random pattern bolting is often employed. A typical pattern would find bolts installed on 5-ft. centers in both directions. To provide further stability and to minimize deterioration from moisture and air, permanent installations may be gunited after bolting.

Beams and roof ties

To increase the bearing area absorbing the load from the bolt, and to control spalling between bolts, steel channels, usually 4-in. at 5.4 lb. per foot or 5-in. at 6.7 lb. per foot, or steel mine roof ties are used in place of mild steel plate washers. Wooden headers are used for the same purpose; however, gradual failure of the wood results in loss of bolt tension, destroying

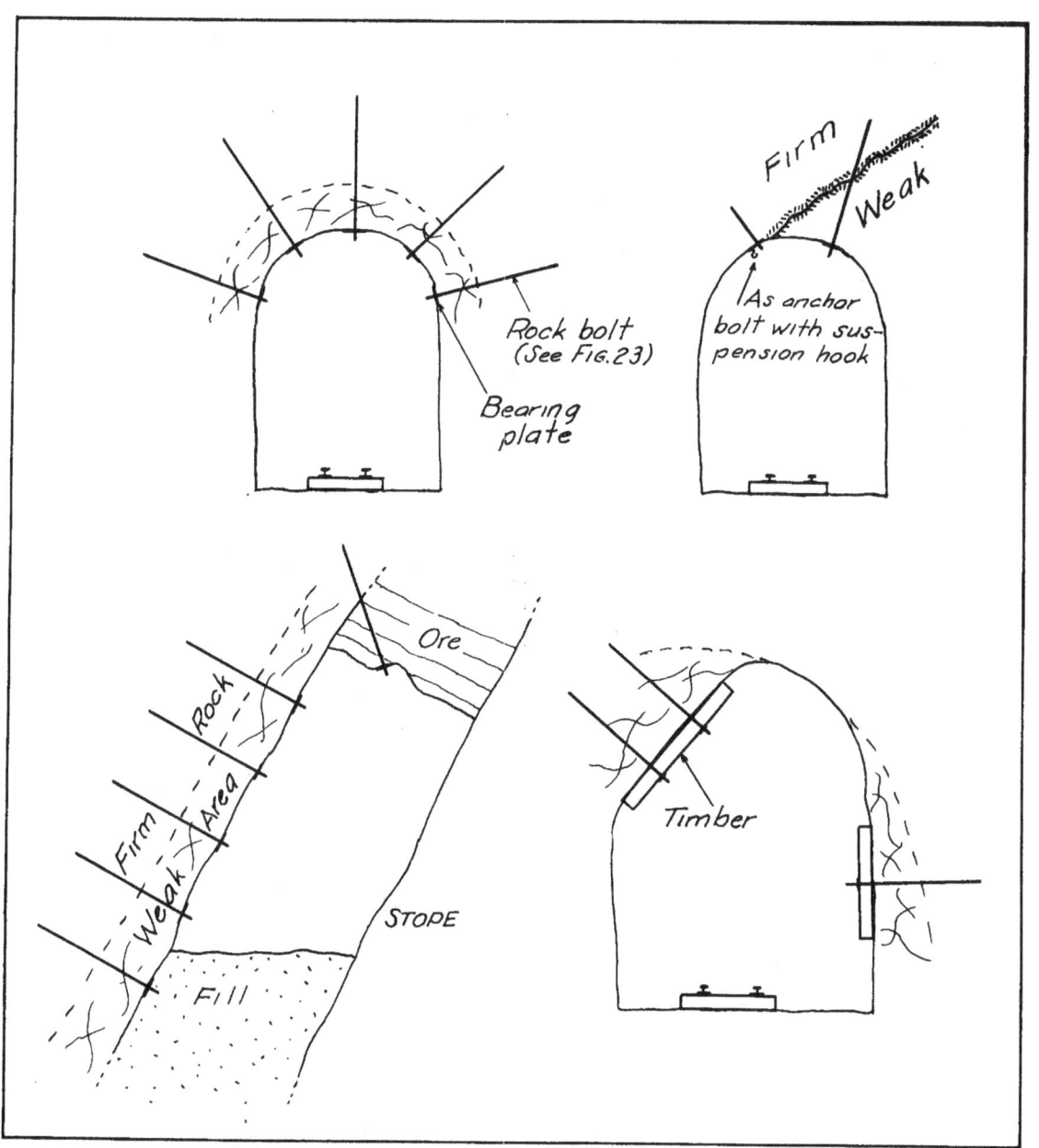

FIGURE 24 — Application of Rock Bolts

part of the effectiveness. Channels and roof ties are provided with holes punched for the spacing of bolts used in the particular formation. Holes are made large enough for either wedge or expansion shell to pass through. Special small plate and angle washers are used to distribute the load from the bolt to the channel or tie. In nonbedded structures they are more difficult to employ in long lengths. Roof ties, on the other hand, are flexible and can be adapted to surface irregularities of almost any ground condition.

Anchorage capacity and bolt loading

The USBM has developed testing procedures to determine the length and type of bolt that should be employed and to measure the possibility of anchoring bolts in any given underground opening. From test results and experimentation, the spacing or pattern necessary to obtain the support desired is determined.

Measurement of anchorage capacity is obtained by pulling installed test bolts with a hydraulic pull tester. The tester is attached to the end of a bolt in the roof with the hydraulic jack bearing against the roof around the bolt, and load is applied until desired values are reached, or until slip occurs. Test results indicate the type and length of bolt that should be used to obtain maximum anchorage for the given rock condition.

Danger of overloading the bolt with the impact wrench used in tightening during installation has led to establishment of safe limits of initial loading. Overloading may not be apparent, and an unsound roof may go unnoticed unless proper procedures are strictly adhered to.

USBM tests have established the ratio of tension in pounds to torque on the nut or bolt head:

a) For 1-in. slotted bolts, tension in lb. = bolt load = 42.5 times torque (foot-pounds) - 1000 lb. (Barry, Panek, and McCormick, 1953, p. 6).

b) For 3/4-in. headed bolts, tension in lb. = bolt load = 39.8 times torque (foot-pounds) (Barry, Panek, and McCormick, 1954, p. 14).

Tests made by Bethlehem Steel Co. indicate wide variation and values in general lower than this unless hardened washers are used under the bolt heads, in which case the tension-torque ratio is 60 for 3/4-in. bolts and 80 for 5/8-in. bolts.

For practical purposes a ratio of 40 to 1 may be used. Tests indicated that manufacturing methods, type of thread lubrication type of bearing plates

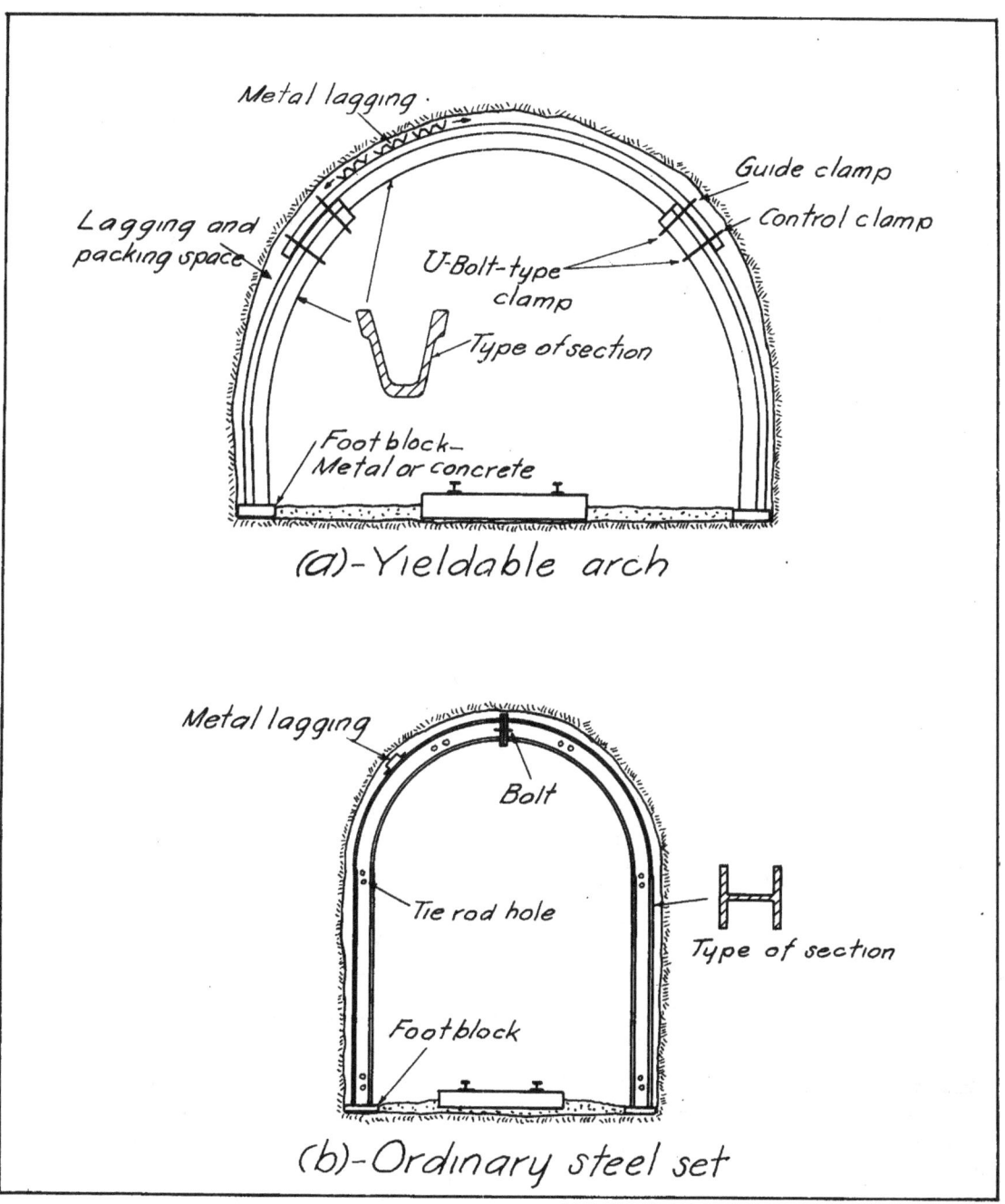

FIGURE 25 – Steel Sets

used, roof strata, and other factors appreciably affect the ratio. For commercial 1-in. bolts the advised limit of installation torque is 350 ft.-lb. or 14,000 lb. Tightening of 3/4-in. headed or expansion shell bolts produces torsional (shear) stresses in addition to tension. Lowering the tensile load at which yield may occur, and a limit of 200 ft.-lb., or 8000 lb. is advised. The torque necessary to turn the nut or the head of the bolt can be read with a dial-type torque wrench. Provided that threads are free from rust and a straight pull is exercised, a true indication of load on the bolt can be obtained. Bolts should be spot checked after installation even though impact wrenches can be set for the desired driving torque. Periodic spot checks should also be made in permanent drifts and accesses to detect any shifting of load. Unless the bolt is in tension, it is not providing adequate support. Loosened bolts should be retightened to insure that rock surface is held in compression.

Problems of bolt use

In addition to torque limitations, adherence to prescribed patterns and the size and depth of hole drilled to install the bolt are of utmost importance. Successful (wedge-type) bolting requires strict adherence to prescribed procedure.

1) A hole of accurate depth. A shallow hole does not allow sufficient takeup on the thread. A hole too deep does not allow proper wedging or tightening. For bolts using 3/8-in. or 1/2-in. plates, or washers, a hole 1 1/2 to 2 in. shorter than the bolt is prescribed.

2) Tight, hard material at the back and sides of the hole. Since the wedge is driven against the back of the hole, it must be sound to provide good anchorage.

3) A hole 1 1/4-in. diameter finished size. A hole too large does not allow enough bite for the legs of the slot when they are wedged out. Thicker wedges may be used if holes are oversize.

For expansion type. Diameter of the hole drilled and protection of bolt threads and expansion shell are of primary importance for proper anchorage of the headed-type bolt. Protection is afforded by shipping and storing bolts with the expansion shell assembled on the bolt. Very good results are obtained where the hole at anchorage depth is of the prescribed size, usually 1 3/8 in. A larger hole reduces both anchorage capacity and effectiveness of the expansion shell. Most shell types are sufficiently strong to break a common steel bolt if the hole is drilled to the proper diameter and the rock structure is firm.

Advantages of roof bolts

Use of roof bolts instead of conventional timber sets offers many advantages, both from the safety and cost standpoint:

1) Safer, more assured roof control. By systematic bolting BEFORE any sag has taken place and by providing sufficient tension in the bolts, roof sag can be limited and roof falls almost eliminated.

2) Lower handling and timbering costs. Reduction of timber in the mine reduces fire hazard.

3) Less storage space required.

4) Cleaner ore and coal through better control of sloughing of waste rock.

5) Reduced danger to men from falls caused by blasting and accidental dislodging of timber posts. Danger from falls in event of train wrecks also lessened.

6) Increased head room and side clearance due to absence of timbers, resulting in increased productivity.

7) Availability of bolts for hanging air, water, and messenger lines.

8) Increased height of working face. Timber sets limit roof heights to approximately 14 ft. before buckling of the legs becomes a problem. Roof bolts allow increased height and, therefore, increased recovery of coal or ore. (One Utah coal mine is now mining the full depth of a seam 25 to 30 ft. thick, removing 80 percent of the coal. Previous recovery limited to only 14 ft. of the seam depth was estimated at 40 to 45 percent).

9) More permanent support. Bolts are estimated to last 20 to 40 years, depending on corrosive conditions.

10) Reduced sloughing of rock caused by attack of air and water. Fractures and faults are kept from opening.

11) Decreased excavation costs. Opening need not be made larger than working space required. In a Butte mine of the Anaconda Co., 2500 ft. of recent development was accomplished with 37 percent

less excavation than would normally be required. Handling 10,000 tons of broken material was avoided through the use of roof bolts.

12) Decreased ventilation costs through reduced resistance to air flow. One mine reports that a single ventilation unit now handles two levels, where one unit per level was required with conventional timbering. Estimated cost per ventilation unit -- $70,000.

Installation sequence

Slotted type roof bolt

1) Hole is drilled with stoper to depth 1 1/2 to 3 in. shorter than bolt to be installed. Finish bit should be 1 1/4-in. in diameter.

2) Back of hole is sounded with rod or bolt to insure soundness.

3) Wedge is inserted into slot.

4) Bearing plate is slipped over threaded end of bolt and nut is screwed on thread so that about 1/2-in. of thread protrudes from nut.

5) Bolt with wedge inserted is started into the hole and threadless dolly fitted onto threaded end of the bolt.

6) Bolt is driven to refusal with stoper. Driving bolt upward forces legs of slot over wedge and into surrounding rock, providing anchorage.

7) Nut is tightened with impact wrench to approximately 350 ft.-lb. torque.

8) Bolt is struck with rod or hammer. Experienced operator can tell by ring if bolt is tightened properly. Spot checks for tightness should later be made with torque wrench.

Wedges used with 1-in. bolts are usually 3/4 x 7/8 x 5 1/2 in. or 6 in. long. Plates are normally 6 x 6 in. or 8 x 8 x 3/8 or 1 1/2 in. thick, with hole centrally punched.

Headed type bolts with expansion shell

1) Hole is drilled slightly longer than the bolt to be installed. Finish bit size should be 1 3/8-in. for most expansion shells.

2) Bearing plate is assembled on bolt and expansion shell threaded on end of bolt.

3) Bolt is inserted in hole and turned to engage shell with sides of hole. As bolt is turned, plug of expansion shell is drawn downward, forcing the expandable sides outward into the rock, providing anchorage.

4) Bolt is tightened with impact wrench and tested by measuring torque with torque wrench.

Among the outstanding articles on rock bolting is that by Thomas and Smedberg (1953-1959, sec. 22, p. 1). Their complete discussion of rock bolting deserves careful reading by anyone expecting to establish a rock-bolting program. This article states that in 1957 over 4,000,000 bolts per month were used in the United States.

USE OF STEEL SETS (FIGURE 25)

While it is unlikely that the small mine operator will encounter ground conditions requiring the use of steel sets, he should know that such sets are available. Fault zones containing heavy, water-saturated gouge are not uncommon. Ground of this kind may best be supported with steel sets*. For designing steel sets see Proctor and White (1946, p. 219-232).

The type of set shown in (b) is available in many varieties: circular rib (shown); elliptical rib; continuous rib separate from the posts and circular (for extremely heavy ground). For spacing between sets see Table 1.

Steel sets are usually blocked in place with wooden blocking. If a concrete lining is to be used, the wooden blocking should be removed. Wood will decay and leave weak sections in the lining.

USE OF CONCRETE SETS (FIGURE 26)

There are localities where timber is not only scarce, but where the timber species available are not suitable for even light ground support. Concrete sets deserve consideration for these localities. In this paper I have made no effort to recommend concrete sets for heavy ground conditions. Such a recommendation would require careful analysis of ground pressures before the set was designed. Here I propose the use of concrete sets only for light ground support when suitable timber is difficult to get. Because this

---
*Commercial Shearing and Stamping Co., Youngstown, Ohio.
Bethlehem Steel Co., Bethlehem, Pa.
Both of the sets shown in Figure 25 are made by these two companies.

substitution of concrete for timber may be primarily an economic problem, the relative costs must be carefully compared.

For proportioning the ingredients of concrete for use under many of the following circumstances, complicated strength and proportioning rules are not necessary: Fuller's rule is sufficient for most set work, as well as for other concrete work for the small operator (footings, walls, floors, foundations).

Fuller's rule for proportioning cement, sand, and gravel is:

$$\text{Cement} = C = \frac{11}{c + s + g}, \text{ bbl. per cu. yd. of concrete.}$$

$$\text{Sand} = C \, s \, \frac{3.8}{27} = \text{cu. yd. of sand per cu. yd. of concrete.}$$

$$\text{Gravel} = C \, g \, \frac{3.8}{27} = \text{cu. yd. of gravel per cu. yd. of concrete}$$

where,

c = proportion of cement.

s = proportion of sand.

g = proportion of gravel.

C = barrels cement per cu. yd. of concrete.

4 sacks cement = one barrel.

This formula is used when the concrete mixture is expressed as a proportion of cement : sand : gravel : (for example, 1:2:4).

Fuller's rule offers a convenient way to express the ingredients of a concrete mixture in easily measured units.

Strength values of 28-day concrete for different mixtures may be taken from Table 4. These values are conservative.

Application of the formula may be shown with a 1:2:4 mixture.

$$\text{Cement} = C = \frac{11}{1+2+4} = \frac{11}{7} = 1.57 \text{ bbl.}$$

$$\text{Sand} = \frac{11}{7} \times 2 \times \frac{3.8}{27} = 0.443 \text{ cu. yd.}$$

$$\text{Gravel} = \frac{11}{7} \times 4 \times \frac{3.8}{27} = 0.886 \text{ cu. yd.}$$

$$C = 1.57 \text{ bbl.} = 1.57 \times 4 = 6.28 \text{ sacks.}$$

To achieve maximum results for a water-cement ratio, consult the Portland Cement Association booklet*. Gallons of water per sack of cement proportions shown in Table 4 will suffice for small batches and such construction requirements for which one usually uses Fuller's rule. The least amount of water consistent with handling the mixture should be used. When necessary, tests may be made using representative samples of the aggregate and water. Final design is based on their results.

In addition, Tables 5, 6, and 7, and Figure 26k give information useful in concrete design.

DESIGN OF CONCRETE SETS

Reinforced concrete design is based on various building and organization-sponsored codes, set up generally for structures directly connected with human safety. Furthermore, structures of this type invariably contain interdependent component parts; if one part fails, the entire project is endangered. In addition, the codes are not entirely uniform from locality to locality. Safety factors used are extremely conservative.

Concrete supports for mines serve a purpose much different from those for ordinary concrete structures. In this discussion concrete is proposed as a substitute for timber when the latter is difficult to obtain in quantity or when mine conditions are conducive to rapid deterioration of timber, or both. Reinforced concrete is unlikely to fail instantly. An unforeseen increase in the loading is invariably shown by progressive cracking of the concrete surface. Only under unusual conditions would the reinforcing steel completely fail to produce instant closure of the drift. Why then, should building code factors of safety be applied to drift sets? I propose that "ultimate strength design" be applied to mine supports (Urquhart, O'Rourke, and Winter, 1958, p. 434; ASCE, 1956). Such a design will save 25 percent (or more) of the materials (weight) in the usual reinforced concrete design. In those mine applications where rock pressures change rapidly and

---

*Design and Control of Concrete Mixtures, 10th Ed., Portland Cement Assn., 33 West Grand Ave., Chicago 10, Ill.

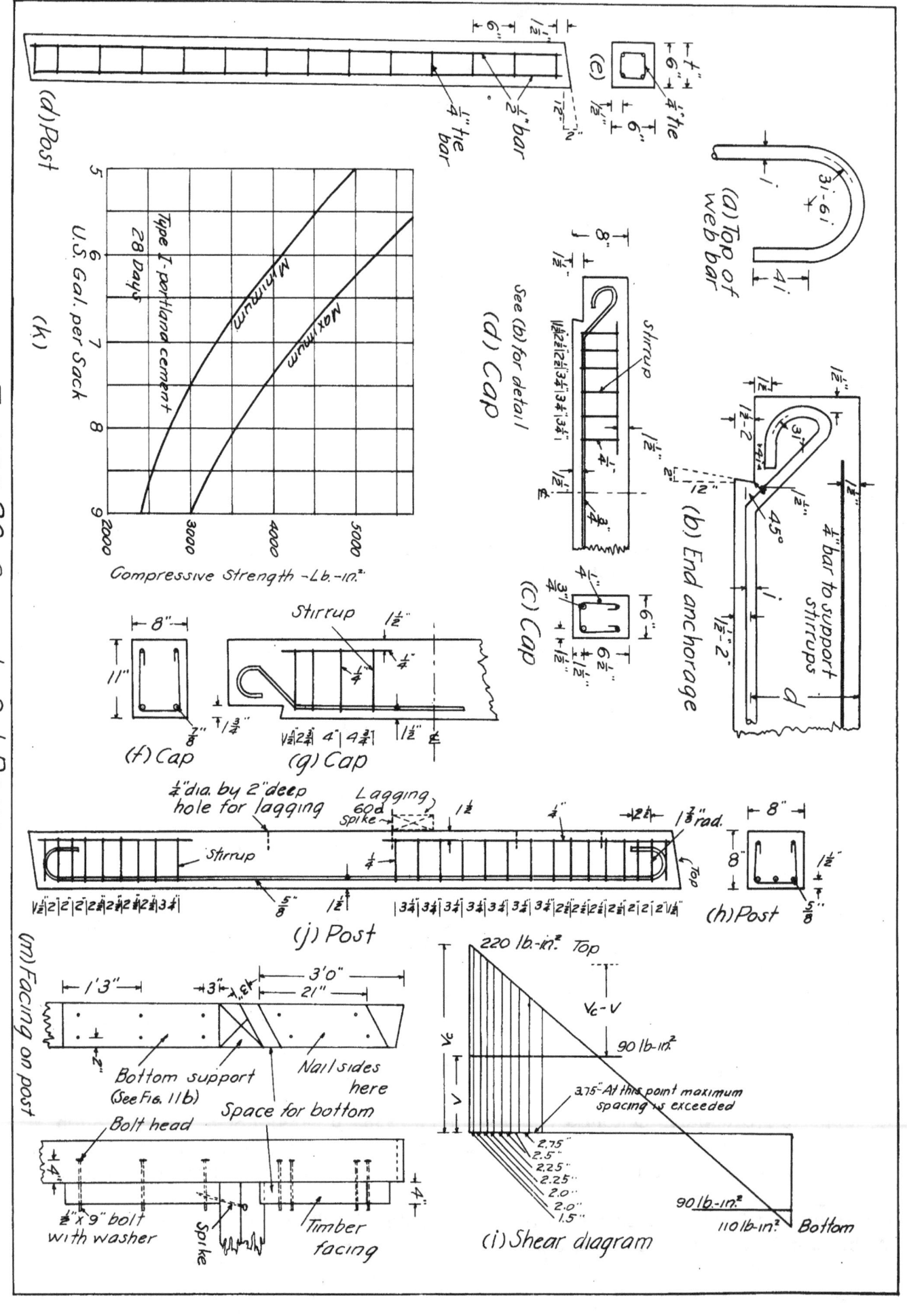

FIGURE 26-Concrete Set Design

are difficult to foretell, the procedure will require modification. In fact, instead of the ultimate strength method it would seem that even a much higher proportion of the 28-day strength for concrete could be used in the ordinary design formulas (40 to 50 percent is the limit recommended in the codes for compression; 75 to 80 percent might well be considered). Similarly for steel, at least the full yield strength should be used. Steel has a much higher ultimate strength than the yield point. For concrete mine-sets a value for steel in excess of the yield point could be considered. In any event, design formulas for concrete sets should be established on their own merits and not tied to building code limits.

When steel is used, costs are based on weight (these include fabricating, transportation, erecting, painting, etc.). When reinforcing bars are selected, weight is an important factor. Several small diameter bars may weigh much less for a given area than fewer large bars of equivalent area.

Two concrete drift sets will be examined:

1) One--an ordinary set--to resist conditions under which an ordinary 8-in. x 8-in. x 5-ft. timber cap would be ample. Here, the load is constant and seldom will increase sufficiently to cause destruction of a set; side pressure is negligible, but scarcity of timber and decay conditions are the reasons for considering concrete.

2) Two--a loaded drift set--to resist the conditions given for the timber set investigated in Figure 5e.

Ordinary set

An 8-in. x 8-in. x 5-ft. timber cap of the species usually found in mining regions would support an evenly distributed total load of about 15,000 lb. This weight is not the ultimate load but one based on the usual safety factors encountered when designing with timber. With this load the maximum bending moment on the cap is 102,400 in.-lb. and the load on each post is about 7600 lb.

Other data required for ultimate strength design are (Urquhart and others, 1958, p. 434; Baker, 1943):

Concrete: f'_c = 3000 lb. per sq. in., 28-day strength; 3/4-in. aggregate with medium sized sand.
n = 10 for 3000-lb. concrete.

Steel: f_y = 40,000 lb. per sq. in. for intermediate grade deformed bars. This grade of steel will be used for all calculations to follow. (See Table 6).

Table 4 -- Strength of Concrete

Water, Gal./Sack of Cement	Mixture Cement:Sand:Gravel	28-day Strength lb./sq. in. (f'_c)	Suggested Allowable Working Stress in Bearing, lb./sq. in. ($0.25 f'_c$)
Less than 5	--	4000	1000
5 - 5 1/2	1:1:2	3000	750
6 - 6 1/2	1:1 1/2:3	2500	625
7 - 7 1/2	1:2:4	2000	500
7 1/2	1:2 1/2:5	1500	375
8	1:3:6	1300	325

Table 5 -- Recommended Slump for Various Uses of Concrete (Parker, 1943, p. 19)

Use	Slump, Inches	
	Minimum	Maximum
Massive sections, pavements, floors laid on ground	1	4
Heavy slabs, beams or walls	3	6
Thin walls and columns, ordinary slabs or beams	4	8

Table 6 -- Areas, Perimeters, and Weights of Standard Steel Reinforcing Bars (Urquhart and others, 1958, p. 502)

Diameter, Inches	Cross-Sectional Area, Sq. In.	Perimeter Inches	Weight per Foot, Lb.
1/4 = 0.250	0.05	0.79	0.167
3/8 = 0.375	0.11	1.18	0.376
1/2 = 0.500	0.20	1.57	0.668

5/8 = 0.625	0.31	1.96	1.043
3/4 = 0.750	0.44	2.36	1.502
7/8 = 0.875	0.60	2.75	2.044
1 = 1.000	0.79	3.14	2.670

Table 7 -- Trial Mixes with Medium Sand and 3-in Slump

Max. Size Aggr., In.	Water, Gal. per Sack	Water, Gal. per Cu. Yd.	Fine Aggr., % Total	Coarse Aggr., % Total
3/4	5	38	45	55
1	5	37	40	60
1 1/2	5	35	36	64
2	5	33	33	67
3/4	5 1/2	38	46	54
1	5 1/2	37	41	59
1 1/2	5 1/2	35	37	63
2	5 1/2	33	34	66
3/4	6	38	47	53
1	6	37	42	58
1 1/2	6	35	38	62
2	6	33	35	65
3/4	6 1/2	38	48	52
1	6 1/2	37	43	57
1 1/2	6 1/2	35	39	61
2	6 1/2	33	36	64
3/4	7	38	49	51
1	7	37	44	56
1 1/2	7	35	40	60
2	7	33	37	63
3/4	7 1/2	38	50	50
1	7 1/2	37	45	55
1 1/2	7 1/2	35	41	59
2	7 1/2	33	38	62
3/4	8	38	51	49
1	8	37	46	54
1 1/2	8	35	42	58
2	8	33	39	61

Bond and Shear: Ordinary design procedures prevail (Report of ASCE, 1956, p. 13).

f_s = 20,000 lb. per sq. in. working stress in steel.

f_c = 0.45f'_c = 1350 lb. per sq. in. working stress for concrete.

f_v = 20,000 lb. per sq. in. working stress in shear for steel.

v = shear = 0.03f'_c = 90 lb. per sq. in. with no web reinforcing.

= 0.08f'_c = 240 lb. per sq. in. with web reinforcing.

u = 0.10f'_c = 300 lb. per sq. in. bond for deformed bars.

Center of steel embedment to concrete surface is 1 1/2 in.

Application of the method requires a trial selection of dimensions of the steel. The ultimate moment or load for this selection is then calculated. A comparison with the maximum moment or load decides whether the section selected is satisfactory.

For this cap the following dimensions will be investigated. (See Figure 26c and d).

To resist shear, these simple beams (caps) will invariably require web reinforcing. Trial dimensions may be found by applying the shear formula,

$$v = \frac{3}{2} \frac{V}{ba}, \text{ lb. per sq. in.}$$

This formula is solved for ba. For v, the value of 240 lb. per sq. in. is used and V = 7600 lb., the post load of the problem.

$$240 = \frac{3}{2} \frac{7600}{ba}$$

ba = 47.5 sq. in.

A 6 x 8 has 48 sq. in. Deducting 1 1/2 in. from the depth of 8 in. gives 6 x 6.5 in. for trial.

b = width = 6 in. ; d = depth to center of steel bar = 6.5 in. ; total depth of section = 6.5 + 1.5 = 8 in.

Two 3/4-in. diameter bars; area one bar = 0.44 sq. in.

Area = A_s = 2 x 0.44 = 0.88 sq. in.

Perimeter = O = 2 x 2.36 = 4.72 in.

$p_{max} = 0.4 \dfrac{f'_c}{f_y} = 0.4 \times \dfrac{3000}{40,000} = 0.03$, maximum reinforcement ratio, must be greater than p.

$p = \dfrac{A_s}{bd} = \dfrac{0.88}{6 \times 6.5} = 0.0226$

$m = \dfrac{f_y}{0.85 f'_c} = \dfrac{40,000}{0.85 \times 3000}$ = maximum steel to maximum concrete ratio.

$T_s = A_s f_y = 0.88 \times 40,000 = 35,200$ lb.

$a = pmd = 0.0226 \times \dfrac{40,000}{0.85 \times 3000} \times 6.5 = 2.30$ in.

$c = d - \dfrac{a}{2} = 6.5 - \dfrac{2.3}{2} = 5.35$ in., internal lever arm -- center of "a" to center of steel

$M_u = T_s c = 35,200 \times 5.35 = 188,300$ in.-lb., resistance of steel section.

Loading originally assumed on the cap was 102,400 in.-lb. The section chosen appears to provide more than sufficient area, although it is about as small as can be used because of shear, bond, and web spacing requirements. Here an inconsistency creeps into the design, because no values for shear and bond correlative to tension and compression have been recommended for ultimate strength design. This inconsistency becomes less serious when one recalls that construction of forms with proper placement of concrete mix becomes difficult as the cross-sectional size of the forms decreases. Dimensions much less than 6 in. x 8 in. will seriously complicate the manufacture of the concrete cap.

<u>Bond, Shear, and Web Spacing</u>

$r = \dfrac{f_s}{f_c} = \dfrac{20,000}{1350} = 14.81$

$k = \dfrac{n}{r+n} = \dfrac{10}{14.81 + 10} = 0.403$

$j = 1 - \dfrac{k}{3} = 1 - \dfrac{0.403}{3} = 0.866$

$$v_c = \frac{V}{bjd} = \frac{7600}{6 \times 0.866 \times 6.5} = 225 \text{ lb. per sq. in.}$$

$$O = \frac{V}{jdu} = \frac{7600}{0.866 \times 6.5 \times 300} = 4.49 \text{ in.}$$

The two 3/4-in bars selected have a perimeter of 2 x 2.36 = 4.72 in.

$$l_1 = \frac{i f_s}{4 u} = \frac{3/4 \times 20,000}{4 \times 300} = 12.5 \text{ in. minimum length of embedment.}$$

As found above, v_c, the actual shear in the section, exceeds v, the allowable shear (90 lb. per sq. in.). Therefore, web reinforcement must be used. The 3/4-in. bars should be extended to form a hook anchorage as shown in Figure 26b. Additional detail for specifying the anchorage is shown in Figure 26.

Web Reinforcement

$$x_c = 1/2 \left(\frac{v_c - v}{v_c}\right) = \frac{54}{2} \left(\frac{225 - 90}{225}\right) = 16.25 \text{ in. from face of support to last stirrup}$$

(1 = length of beam (cap), center of post to center of post = 54 in.)

$$V_s = 1/2 (v_c - v) x_c b = 1/2 (225 - 90) \times 16.25 \times 6 = 6560 \text{ lb.}$$

For web reinforcing, the diameter of the bars should approach 1/50 of d and the web bars be otherwise dimensioned as shown in Figure 26a.

Because 1/4-in. diameter bars are the smallest obtainable, they must necessarily be used. There are two sides to the stirrup; hence, twice the area of one bar is used.

$$A_v = 2 \times 0.05 = 0.10 \text{ sq. in.}$$

$$N_v = \frac{V_s}{f_v A_v} = \frac{6560}{20,000 \times 0.10} = 3.28 \text{ or 4 stirrups.}$$

$$s = \frac{f_v A_v}{(v_c - v) b} = \frac{20,000 \times 0.10}{(225 - 90) \times 6} = 2.47 \text{ or 2.5 in. minimum spacing.}$$

Maximum spacing = d/2 = 6.5/2 = 3.25 in.

Baker (1943, part 2, p. 31) suggests grouping the stirrups as follows:

1) About 1 to 2 in. from the face of the support for the first stirrup.

2) At $\dfrac{x_c}{4}$, spacing between stirrups is $\dfrac{4s}{3}$ for first group.

3) At $\dfrac{x_c}{2}$, spacing between stirrups is 2s for second group.

4) At $\dfrac{3x_c}{4}$, spacing between stirrups is 4s for third group.

5) Last stirrup at x_c, with spacing in 4th group not to exceed d/2.

6) Spacing not to exceed d/2 at any position.

These spacings are only approximate, and N_v is the minimum number required. Additional stirrups may be used to balance the grouping. Because of the small cross-sections generally designed for mine support sets (compared to building construction sets), the stirrups occur closely spaced. In choosing between the minimum and maximum spacing, one may have to decide on the basis of convenience in placing the concrete.

After the above six suggestions were applied, the spacing and number of stirrups shown in Figure 26d (cap) were decided upon.

For the set under consideration very little pressure from ground movement is assumed to be acting against the posts. If the spreader is 1 1/2 in. deep, it and the post should resist;

6 x 1 1/2 x 750 lb. per sq. in. (see Table 4) = 6750 lb.

Post

The given axial load on each post is 7600 lb. One dimension should correspond to the width of the cap (6 in.).

For a trial, a 6-in. x 6 in. section (Figure 26e) will be investigated. At each corner 1/2-in. diameter bars tied together will be used. This arrangement is known as rectangular reinforcing. (Spiral reinforcing is usually applied to round columns). In designing columns (posts), the total dimensions are used. Here the depth equals t or 6 in. Also, (Urquhart and others, 1958, p. 452) an eccentric loading is assumed. When an average eccentric load is used with this type of column, the eccentricity e' is taken as 0.1t. Values for the concrete and steel are the same as those used for the cap.

Column design also depends on whether the member is a long or a short column. Short columns are those whose length does not exceed 15 times the least dimension.

For the post under consideration the maximum figure would be 15 x 6 in. or 90 in. Actually the post will not exceed about 6 1/2 feet or 78 in. When the member is a long column (Urquhart and others, 1958, p. 459),

the allowable load = $P(1.6 - \frac{0.04L}{t})$, where L = unsupported length in inches, and t = least dimension of the column.

This formula may also be expressed as

$$\frac{P}{(1.6 - \frac{0.04L}{t})} = \text{the load for which a long column must be designed.}$$

Load on post is 7600 lb.; b = 6 in. and t = 6 in.

$$P_u = \frac{2 A'_s f_y}{\frac{2e'}{(d-d')} + 1} + \frac{b t f'_c}{\frac{3te'}{d^2} + 1.18}, \text{ lb.}$$

d = distance from the extreme face to the center of the reinforcing bar in the direction t.

d' = distance from the surface of the concrete to the center of the steel (actually the embedment distance).

A'_s = area of the steel bars for compression.

Substituting values,

A'_s = 4 x 0.20 = 0.80 sq. in. (1/2-in. bars used).

d' = 1 1/2 in.

d = 6 in. - 1 1/2 in. = 4 1/2 in.

e' = 0.1 x 6 = 0.6 in.

$$P_u = \frac{2 \times 0.80 \times 40,000}{\frac{2 \times 0.6}{(4\ 1/2 - 1\ 1/2)} + 1} + \frac{6 \times 6 \times 3000}{\frac{3 \times 6 \times 0.6}{4.5^2} + 1.18}$$

+ 45,700 + 63,200 = 108,900 lb.

Regardless of this great excess in capacity, the section is about the smallest size practical. Anything smaller would, among other things, prevent proper distribution of the aggregate around the bars. This is not an unusual example of the selection of a member for reasons other than the stresses to be resisted.

To help hold the four longitudinal bars in position, ties will be used. Spacing between the ties is the minimum distance derived from either

        16 x diameter of reinforcing bars;
        48 x diameter of ties;

or the least dimension of the column.

Material used for the ties should have a minimum diameter of 1/4-in. For the above column, 1/4-in. diameter bars will be used for the ties.

        16 x 1/2-in = 8 in.
        48 x 1/4-in. = 12 in.
        Least dimension of the column = 6 in.

Therefore, the spacing will be 6 in. from center to center of the ties. Bearing between the posts and the cap is,

6 x 6 x 750 lb. per sq. in. = 27,000 lb., which is more than ample. Detail for the post is shown in Figure 26d (post).

Loaded drift set -- cap

Data used for designing the timber set in Figure 5e were used for finding a concrete set capable of resisting the same conditions. Calculations are identical to those just presented. Values for the steel and concrete are the same as assumed for the ordinary set.

Figure 26f shows the section and dimensions selected. For reasons similar to those given for the previously designed cap (vertical shear at the post, bond, etc.), the section is somewhat larger than would be necessary for tension and compression alone.

Figure 26g shows the stirrup spacing and end anchorage.

Post

Because this post may have a large bending stress, the design must consider both bending and axial load.

Loading to be resisted by the post is,

    $M = 170,800$ in.-lb. in bending $= (M_u)$.

P = 12,000 lb. axial load (includes weight of cap).

The first step is to determine whether compression or tension governs the design. Compression is first checked.

$$P_u = 12,000 \text{ lb.} = \text{axial load on column.}$$

When P_b is greater than P_u, failure will occur from tension, and the member is designed as a beam.

$$P_b = (0.72 \frac{90,000}{f_y + 90,000} f'_c) + A'_s - A_s f_y, \text{ lb.}$$

A'_s is area of steel in compression; for the column under examination, no steel is used in the compression side of the column (side next to wall rock).

Figure 26h shows the post section.

$$b = 8 \text{ in.}; \quad d = 6.5 \text{ in.}; \quad \text{and } t = 8 \text{ in.}$$

(A preliminary check for the bond resistance indicates that the three 5/8-in. bars are suitable).

$$A_s = 3 \times 0.31 = 0.93 \text{ sq. in.}$$

$$O = 3 \times 1.96 \text{ in.} = 5.88 \text{ in.}$$

$$P_b = (0.72 \times \frac{90,000}{40,000 + 90,000} \times 3000 \times 8 \times 6.5) + 0 - 0.93 \times 40,000$$

$$= 77,600 - 37,200 = 40,400 \text{ lb. which is greater than}$$

$$P_u = 12,000 \text{ lb.}$$

Therefore, tension governs and the section must be designed as a beam.

$$M_u = C_c \left(\frac{t}{2} - \frac{a}{2}\right) + A_s f_y \left(d - \frac{t}{2}\right), \text{ in.-lb.}$$

$$C_c = 0.85 f'_c b a, \text{ lb.}$$

$$a = \frac{P_u + A_s f_y}{0.85 f'_c b}, \text{ in.}$$

$$a = \frac{12{,}000 + 0.93 \times 40{,}000}{0.85 \times 3000 \times 8} = 2.19 \text{ in.}$$

$$M_u = 0.85 \times 3000 \times 8 \times 2.19 \left(\frac{8}{2} - \frac{2.19}{2}\right) + 0.93 \times 40{,}000 \left(6.5 - \frac{8}{2}\right)$$

$$= 130{,}000 + 93{,}000 = 223{,}000 \text{ in.-lb.}$$

Section chosen is satisfactory.

Spreader Bearing

$8 \times 1\,3/4 \times 750 = 10{,}500$ lb., which is greater than 9933 lb.

Bond, Shear, and Stirrup Spacing

j has same value as before.

$$v = \frac{V}{bjd} = \frac{9933}{8 \times 0.866 \times 6.5} = 220 \text{ lb. per sq. in., top of post.}$$

$$v = \frac{4967}{8 \times 0.866 \times 6.5} = 110 \text{ lb. per sq. in., bottom of post.}$$

$$O = \frac{V}{jdu} = \frac{9933}{0.866 \times 6.5 \times 300} = 5.86 \text{ in., which checks the three 5/8-in. bars.}$$

For web reinforcing, 1/4-in. bars will be used,

$$A_v = 2 \times 0.05 = 0.10 \text{ sq. in.}$$

$$s_c = \frac{f_v A_v}{(v_c - v) b} = \frac{20{,}000 \times 0.10}{(220 - 90) \times 8} = 1.92 \text{ or 2 in. minimum spacing.}$$

Maximum spacing = $d/2 = 6.5/2 = 3\,1/4$ in.

Because of the unsymmetrical loading on the post (load increases from zero at the bottom of the post to a maximum at the top), the spacing of the stirrups is best found by using a shear diagram. (See Figure 26i).

From,

$$s = \frac{f_v A_v}{(v_c - v) b}, \text{ in.,}$$

we note that $\dfrac{f_v A_v}{b}$ is a constant $\left(\dfrac{20{,}000 \times 0.10}{8} = 250\right)$. Values of $(v_c - v)$ may be scaled from the shear diagram. Dividing these scaled values into the constant gives the spacing between segments. Stirrups are placed at the center of each segment.

It must be kept in mind that the maximum is still d/2 inches. In the post under consideration the spacing is between 2 in. and 3 1/4 in. The first stirrup will be placed 1 1/2 in. from the top of the post as shown in Figure 26j. Succeeding stirrups will be spaced as shown.

There is a little shear in excess of v at the bottom of the post. Merely as a precautionary measure, stirrups are placed there as shown.

Ordinary set (Customary method of design)

To show the comparison in size and weight, the cap for the ordinary set will be selected.

Following this procedure for the design, the values used for the concrete and steel are,

$$f_c = 0.45 f'_c = 0.45 \times 3000 = 1350 \text{ lb. per sq. in.}$$

When intermediate steel deformed bars are used,

$$f_s = 20,000 \text{ lb. per sq. in.}$$

$$v = 90 \text{ lb. per sq. in.}; \quad n = 10; \quad j = 0.866; \quad k = 0.403.$$

For this procedure the design is based on the balanced section method.

$$bd^2 = \frac{2M}{f_c k j}, \text{ in.}^3 \text{ and } A_s = \frac{M}{f_s j d}, \text{ sq. in.}$$

Members of these formulas have the same designation as before.

As in the ultimate strength design, the cap must be checked for vertical shear at the face of the post.

$$v = \frac{3}{2} \frac{V}{ba}, \text{ lb. per sq. in.}$$

With web reinforcing, v is equal to 240 lb. per sq. in.

$$240 = 3/2 \times \frac{7600}{ba} \text{ and}$$

$$ba = 47.5 \text{ sq. in.}$$

If b = 6 in. and a = 8 in., sufficient area for the shear is available. For this selection d will be assumed at 8 1/2 in. If the embedment is 1 1/2 in., the total depth becomes 10 in. Two 3/4-in. deformed bars are used.

$$M = \frac{bd^2 f_c kj}{2} \text{, in.-lb.}$$

$$= \frac{6 \times 8.5^2 \times 1350 \times 0.403 \times 0.866}{2} = 102,300 \text{ in.-lb.}$$

$$A_s = \frac{102,300}{20,000 \times 0.866 \times 8.5} = 0.677 \text{ sq. in.}$$

Two 3/4-in. bars = 2 x 0.44 = 0.88 sq. in.

O = 2 x 2.36 = 4.77 in.

or $O = \frac{7600}{0.866 \times 8.5 \times 300} = 3.45$ in.

Cap is 6 in. x 10 in. x 5 ft. long.

Weight = $\frac{6 \times 10}{144}$ x 150 lb. per sq. in. x 5 ft. = 313 lb.

Weight of previous design = $\frac{6 \times 8}{144}$ x 150 x 5 = 250 lb.

Saving in weight is 313 - 250 = 63 lb. so far as concrete is concerned.

Calculation of concrete mix

If an occasional, small amount of concrete is needed and the exact strength is of little importance, Fuller's rule will suffice. On the other hand, if concrete in quantity with a specified strength is desired, a more exact method for figuring the mix should be used, as would be true if a considerable number of sets were to be made. To make such calculations consult data for concrete mixtures. Design and Control of Concrete Mixtures (1952, p. 18) is recommended. Sets should be allowed to cure for at least 28 days. During this period the surface of the concrete should be kept well moistened.

To illustrate the calculations, we shall determine a mix for the concrete sets previously investigated.

Raw materials (sand, aggregate, and water) will vary widely. Care must be exercised to use materials free from dirt, clay, and organic substances. Absolutely dry sand or aggregate will seldom be available. Therefore, the percent of moisture in each must be closely estimated. Of the five types of portland cement, only the normal or common type will be considered--this type will probably always be available and is used for most ordinary construction.

In the calculations for the sets a 3000-lb. per sq. in. concrete was used. To make this concrete, assume the following data:

1) Medium sand, 2 percent moisture. Medium here refers to the average size of the sand grains, which is defined and explained in Design and Control of Concrete Mixtures (1952).

2) Maximum size of aggregate, 3/4-in. with 1 percent moisture.

3) Water should be determined on the basis of a 15 percent stronger concrete (Design and Control of Concrete Mix., 1952, p. 7): 3000 x 1.15 = 3450 lb. per sq. in.

4) From tabulated information (Design and Control of Concrete Mix, 1952, p. 5 and 16; Fig. 26k is from this publication), a 3450-lb. concrete requires 6 3/4 to 8 1/4 gal. of water per sack (94 lb.) of cement; nearest trial amount is 7 gal. For the 3/4-in. aggregate, medium sand, and 7 gal. per sack, 49 percent of sand, 51 percent of aggregate, and 38 gal. of water are necessary per cubic yard. (See Table 7, extracted from Table 5, Design and Control of Concrete Mix.).

5) Slump of concrete mix is to be 5 in. (See Table 5), to provide sufficient fluidity for filling the forms.

6) Correction for slump (tabulated data figured on slump of 3 in.); for each 1-in. change in slump, increase or decrease water by 3 percent.

$$38 \text{ gal.} \times 1.06 = 40 \text{ gal. for 5-in slump.}$$

7) Specific gravity of cement, 3.15; of sand and aggregate, 2.65.

8) Cement factor = $\frac{40}{7}$ = 5.72 sacks per cu. yd.

<u>For one cubic yard</u>

Absolute volume of cement = $\frac{94 \times 5.71}{3.15 \times 62.4}$ = 2.73 cu. ft.

Volume of water = $\frac{40}{7.48}$ = 5.35 cu. ft.

Volume of cement-water = 8.08 cu. ft.

Absolute volume of aggregate = 27.00 - 8.08 = 18.92 cu. ft.

Absolute volume of sand = 0.49 x 18.92 = 9.26 cu. ft.

Weight of surface-dry sand = 9.26 x 2.65 x 62.4 = 1531 lb.

Absolute volume of aggregate = 0.51 x 18.92 = 9.65 cu. ft.

Weight of surface-dry aggregate = 9.65 x 2.65 x 62.4 = 1598 lb.

For each sack of cement:

Weight of surface-dry sand = $\dfrac{1531}{5.71}$ = 268 lb.

Weight of surface-dry aggregate = $\dfrac{1598}{5.71}$ = 280 lb.

Correction for moisture:

Free moisture in sand = 268 x 2% = 5.36 lb.

Free moisture in aggregate = 280 x 1% = 2.80 lb.

Total free moisture = $\dfrac{5.36 + 2.80}{8.33}$ = 0.98 = 1 gal.

Weight of moist sand = 268 + 5 = 273 lb.

Weight of moist aggregate = 280 + 3 = 283 lb.

Water to be added = 7 gal. - 1 gal. = 6 gal.

Mix to be used: 94 lb. cement, 273 lb. sand, 283 lb. aggregate, and 6.0 gal. of water.

To make one cubic yard of concrete mix:

Cement: 94 lb. x 5.71 sacks	=	537 lb.
Sand: 273 x 5.71	=	1560 lb.
Aggregate: 283 x 5.71	=	1615 lb.
Water: 6 gal. x 5.71	=	34.2 gal.

For very important work, the preceding calculations would be used for a test batch. Adjustment of succeeding batches would follow until the required strength was obtained.

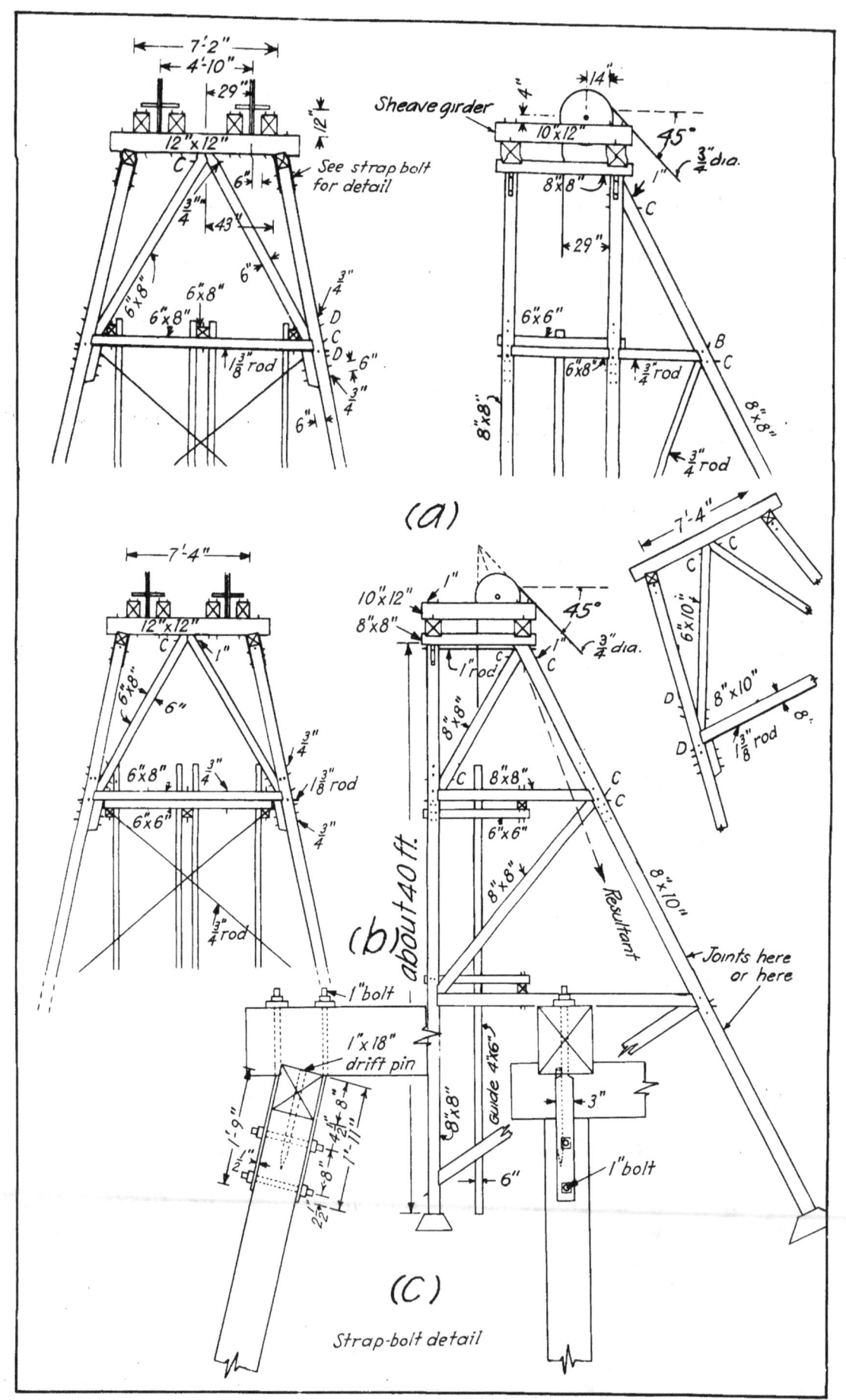

FIGURE 27—Wooden Headframe

Placing sets in drift

To insure a tight, even-bearing contact between the top of the post and the cap, a thin layer of thick grouting should be used. A quick setting type of cement is best.

Chute sets

Concrete sets must be designed to permit the construction of chutes between sets. Figure 26m suggests a design for facing the posts with timber. (Wooden sets could be used instead of concrete when chutes are required). To provide ample thickness for spiking the chute to the set, 4-in. thick material is recommended. When spiking to the facing, incline spikes to the surface; thus, if the spikes pass clear through the timber facing, they will tend to bend at the concrete face instead of at the spike-head.

Some of the plastic adhesives on the market might be investigated for fastening the wooden facing to the concrete.

Side lagging

To support the side lagging, 5-in., 1/4-in. diameter rods or large spikes (say 40d) may be embedded 2 inches in the back of the post. Spacing is selected for the conditions under consideration. Lagging simply rests on the 3-in. projection of the bolt or spike. It is not necessary to bolt the boards in place.

Instead of casting these bolts integral with the post, cast a hole of sufficient diameter to receive the bolt or spike, and about 2 in. deep in the concrete (See Figure 26j). Spikes may then be inserted in these holes to support the lagging after the set is erected.

WOODEN HEADFRAMES (FIGURE 27)

Two designs for a wooden headframe are shown in the figure. Additional information may be found in Staley (1949, p. 105-186); (1937, p. 1-37); and Tillson (1938, p. 30-48).

For the construction shown, the use of solid material is assumed. However, the construction may be built up (laminated) with planks. If the headframe is designed in this way, an additional two inches should be added to the dimensions shown for the solid members. The right number of two-inch planks to give width and thickness will be used. Odd lengths are used so that there will be overlapping. Not only is the assembly well spiked together but a pair of 3/4-in. bolts are used every 30 inches. Center-of-bolt holes are spaced 2 inches from the edges, and set in 3 inches from the ends. Planks for laminated construction must be carefully selected to be free of knots, checking, and crossgrain.

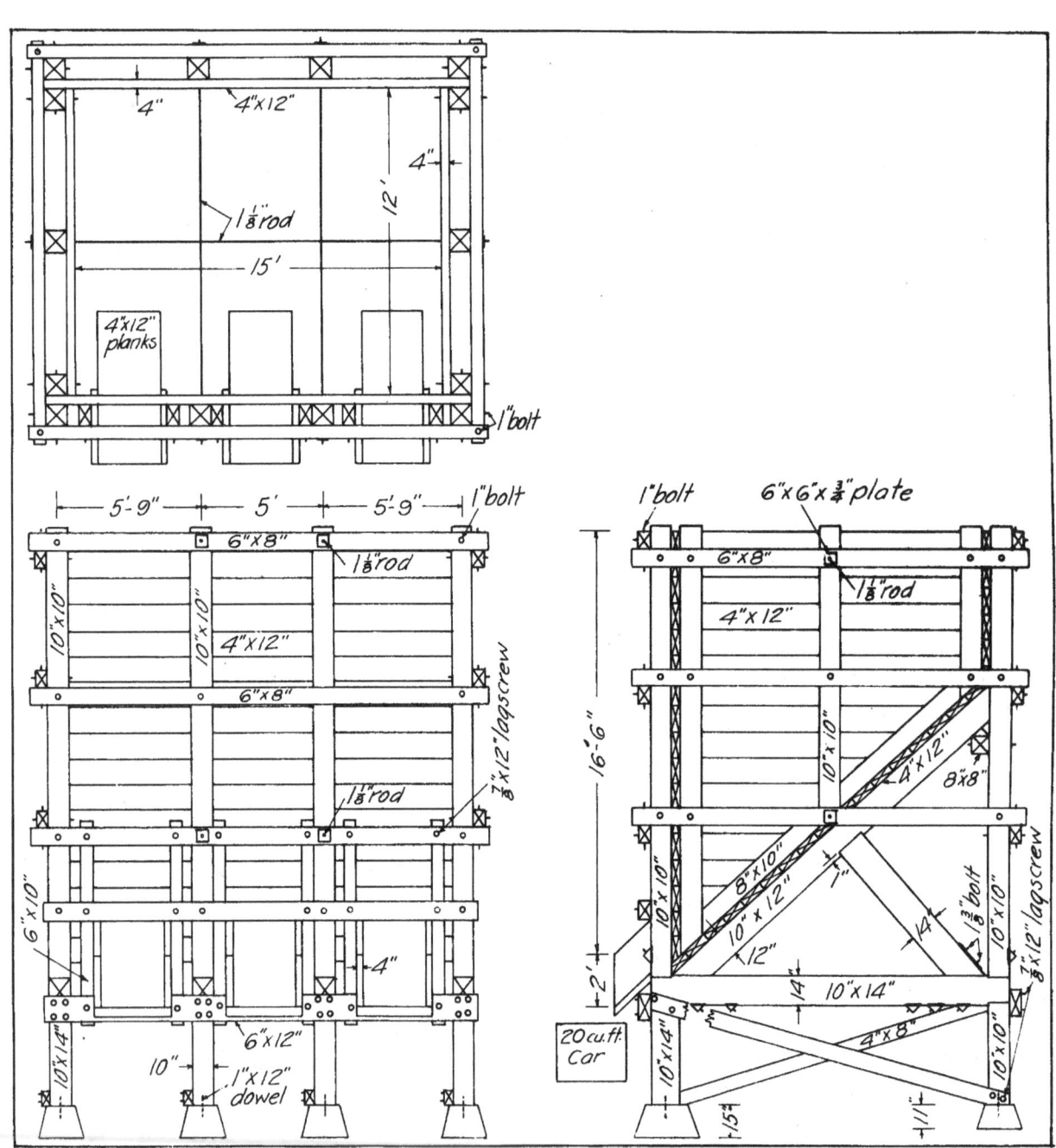

FIGURE 28—Wooden Ore Bin.

The structures shown were designed for a load of 5000 lb., which includes cage, car, ore, and rope. Acceleration or other forces were not included.

If a headframe higher than the 40 ft. shown is necessary, the design given may be altered as long as the height is divided into sections not exceeding 10 to 12 ft. each. The overall height should not exceed 60 ft. unless the stresses in the struts and bracing are checked.

The distance between the back posts and the front posts should always be such that the resultant stress between the vertical and inclined portion of the rope falls well within the back post, as shown in Figure 27.

Table 8 gives size of solid timbers for main members if a rope pull greater than 5000 lb. is desired (Staley, 1937). This table is based on the assumption that the dry timber has allowable working stresses similar to any species of Douglas fir, hemlock, tamarack, or longleaf southern yellow pine. For species with lower working stresses, the next larger commercial size is suggested (for example, an 8 x 10 would be increased to 10 x 12).

Guides for the cage should be made of the best grade of selected Douglas fir or longleaf southern yellow pine. Joints for guides should occur at the shaft timbers. For making the connection, the half-lap scarf joint should be used. All bolts or lagscrews used on the face of the guide (contact with cage guide shoes) should be countersunk well below the surface of the guide.

Splices for posts and struts may be butt joints with a splice plate on each side, or the half-lap scarf joint may be used. Splices in the posts should be as near the panel joint as convenient.

For details on timber joints consult the Wood Handbook (1935, p. 119-136).

WOODEN ORE BIN (FIGURE 28)

Figure 28 shows an ore bin designed for a capacity of 100 tons. The lumber for this bin is assumed to be common grade of fir (nearly everything other than pine and cedar seems to be called fir), tamarack, or carefully selected grades of pine. If greater capacity is desired, the bin may be lengthened by additional chute sections. Further information on ore bins may be obtained from Staley (1949, p. 187-211) or Taggart (1945, sec. 18, p. 1-19).

Concrete footings for bins are preferable to timber sills. For the bin shown the dimensions of the footings are:

Table 8 -- Essential Details for Various Loads on Wooden Headframes

Rope pull, lb. (cage, car, ore, rope)	Rope dia., in.	Breaking strength of rope, lb.	Size of posts, Inches		Struts, in.	Bracing, dia, rod, in.	Sheave groove dia., in.	Sheave girder, in.
			Front	Back				
4,500	5/8	33,200	8 x 8	8 x 8	4 x 8	3/4	30	10 x 12
6,500	3/4	47,400	8 x 10	8 x 10	6 x 10	3/4	36	10 x 12
8,500	1	84,000	8 x 12	12 x 12	6 x 12	1	48	12 x 12
10,000	1	84,000	8 x 12	12 x 12	6 x 12	1	48	12 x 12

Table 8 -- (Continued) Footings, Concrete, Inches

Front posts			Back posts			
Top	Bottom	Depth	Top	Bottom	Depth	Base Plate
11 1/2	21	11	14	36	22	10 x 10 x 1 3/8
12 1/2	24	14	17	45	32	11 x 13 x 1 1/2
17	32	18	22	59	38	15 1/2 x 15 1/2 x 1 1/2
17	32	18	22	59	38	15 1/2 x 15 1/2 x 1 1/2

Front Posts--

    Top: 1 ft. 8 in. by 1 ft. 2 in.

    Bottom: 2 ft. 7 1/2 in. by 1 ft. 10 1/2 in.

    Depth: 1 ft. 3 in.

Back Posts--

    Top: 1 ft. 2 in. by 1 ft. 2 in.

    Bottom: 1 ft. 10 1/2 in. by 1 ft. 10 1/2 in.

    Depth: 11 in.

REFERENCES CITED

Baker, Samuel, 1943, Reinforced-concrete design: International Textbook Co., Parts 1, 2, and 3, Scranton, Pa.

Barry, A. J., Panek, L. A. and McCormick, J. A., 1953, Use of torque wrenches to determine load in roof bolts; Part I - Slotted type bolts: U. S. Bur. Mines R. I. 4967, 7 p.

_____, 1954, Use of torque wrenches to determine load in roof bolts; Part II - Expansion type 3/4-in. bolts: U. S. Bur. Mines R. I. 5080, 17 p.

Basic Estimating, (-----), International Harvester Co., Melrose Park, Ill., 74 p.

Boyd, J. E., 1917, Strength of materials, 2nd Ed.: McGraw-Hill Book Co., Inc., New York.

Chemical injections ease tunneling through wet, sandy material, Nov. 1958, Eng. and Mng. Jour., New York.

Design and control of concrete mixtures, 10th Ed., 1952, Portland Cement Assn., 33 West Grand Ave., Chicago 10, Ill., 68 p.

Farmin, Rollin and Sparks, C. E., Sept. 1953, The use of wooden rock bolts in the Day Mines; Mining Eng., 3 p.

Feather River and Delta Diversion Project, 1959, Appendix C - Procedure for estimating costs of tunnel construction: State of California Dept. of Water Resources Bull. No. 78, Sacramento, California, 73 p.

Gardner, E. D. and Vanderburg, W. O., 1933, Square-set system of mining: U. S. Bur. Mines I. C. 6691, 73 p. (out-of-print).

Humphrey, J. L., May 1956, Steel bolts in mine roof support: Mining Eng., 5 p.

In Sweden, the method is simple, Sept. 1956, Eng. and Mng. Jour.

Mining Laws of the State of Idaho, 1959, State Mine Inspector, Boise, Idaho, 120 p.

Parker, Harry, 1943, Simplified design of reinforced concrete: John Wiley and Sons, Inc., New York.

Peele, Robt., (Ed.), 1941, Mining Engineers' Handbook: John Wiley and Sons, Inc., New York.

Proctor, R. V. and White, T. L., 1946, Rock tunneling with steel supports: The Commercial Shearing and Stamping Co., Youngstown, Ohio, 278 p.

Pynnonen, R. O. and Look, A. D., 1958, Chemical solidification of soil in tunneling at a Minnesota iron-ore mine: U. S. Bur. Mines I. C. 7846, 8 p.

Report of ASCE - ACI Joint Committee, 1956, Ultimate strength design: Proc. Amer. Soc. Civ. Eng., v. 81, paper no. 809, New York 18, 68 p.

Staley, W. W., 1937, Design of small wooden headframes: U. S. Bur. Mines I. C. 6843, 37 p. and 16 fig. (out-of-print).

_____, 1949, Mine plant design 2nd Ed, McGraw-Hill Book Co., Inc., New York, 540 p.

_____, 1961, Prospecting and developing a small mine: Idaho Bur. of Min. and Geol. Bull 20, 104 p.

Sun, S. C. and Purcell, G., 1959, The importance of electrokinetics in mineral flotation: Mineral Industries v. 29, Oct. 1959, The Pennsylvania State College, University Park, Pa.

Taggart, A. F., 1945, Handbook of mineral dressing: John Wiley and Sons, Inc., New York.

Thomas, E. M. and Smedberg, M., 1953-1959, Manual on rock blasting - Rock bolting sec. 22: Atlas Diesel Co., Stockholm, Sweden, 58 p. (Available in English, Swedish, French, German. May be obtained through Atlas Copco Inc., 545 Fifth Ave., New York 17. There are 22 sections on as many different subjects related to rock blasting).

Tillson, B. F. 1938, Mine plant: Amer. Inst. of Min. and Met. Eng., New York, 371 p. (out-of-print).

Urquhart, L. C., O'Rourke, C. E., and Winter, Geo., 1958, Design of concrete structures 6th Ed. : McGraw-Hill Book Co., Inc., New York, 546 p.

Wood Handbook, 1935, Forest Products Laboratory, U. S. Dept. Agr., 325 p.

www.ingramcontent.com/pod-product-compliance
Lightning Source LLC
Chambersburg PA
CBHW081836170526
45167CB00007B/2832